AF902106

Emerging Trends in Construction Information Management

Peter Adekunle · Clinton Aigbavboa ·
Opeoluwa Akinradewo · Ayodeji Oke

Emerging Trends in Construction Information Management

Springer

Peter Adekunle
Department of Construction Management
and Quantity Surveying
University of Johannesburg
Johannesburg, South Africa

Clinton Aigbavboa
Department of Construction Management
and Quantity Surveying
University of Johannesburg
Johannesburg, South Africa

Opeoluwa Akinradewo
Department of Quantity Surveying
and Construction Management
University of the Free State
Bloemfontein, South Africa

Ayodeji Oke
Department of Quantity Surveying
Federal University of Technology
Akure, Nigeria

ISBN 978-3-032-09678-4 ISBN 978-3-032-09679-1 (eBook)
https://doi.org/10.1007/978-3-032-09679-1

© The Editor(s) (if applicable) and The Author(s), under exclusive license to Springer Nature Switzerland AG 2026

This work is subject to copyright. All rights are solely and exclusively licensed by the Publisher, whether the whole or part of the material is concerned, specifically the rights of translation, reprinting, reuse of illustrations, recitation, broadcasting, reproduction on microfilms or in any other physical way, and transmission or information storage and retrieval, electronic adaptation, computer software, or by similar or dissimilar methodology now known or hereafter developed.
The use of general descriptive names, registered names, trademarks, service marks, etc. in this publication does not imply, even in the absence of a specific statement, that such names are exempt from the relevant protective laws and regulations and therefore free for general use.
The publisher, the authors and the editors are safe to assume that the advice and information in this book are believed to be true and accurate at the date of publication. Neither the publisher nor the authors or the editors give a warranty, expressed or implied, with respect to the material contained herein or for any errors or omissions that may have been made. The publisher remains neutral with regard to jurisdictional claims in published maps and institutional affiliations.

This Springer imprint is published by the registered company Springer Nature Switzerland AG
The registered company address is: Gewerbestrasse 11, 6330 Cham, Switzerland

If disposing of this product, please recycle the paper.

Competing Interests The authors have no competing interests to declare that are relevant to the content of this manuscript.

Contents

Chapter 1
Introduction

Abstract Construction information management (CIM) is essential to the effective completion of construction projects because it makes sure that all parties involved have access to timely, accurate, and pertinent data. Managing enormous volumes of data from design documents to site reports and material specifications has become crucial to project efficiency, safety, and cost control as the construction industry grows more complicated. To prepare for a more thorough examination of CIM's methods, resources, and practical applications later in the book, this chapter presents the core ideas and importance of the field. The context of the study is presented, emphasising the increasing need for integrated communication systems and digital transformation in the construction industry, particularly in response to global issues including resource shortages, urbanisation, and environmental imperatives. The chapter also describes the purpose and goals of the study, highlighting its emphasis on enhancing decision-making, data accuracy, and collaboration using efficient information management techniques. It also talks about the work's worth and impact on both industry practice and scholarly publications. The book's format is briefly explained, providing readers with a clear road map for navigating the case studies and issues discussed. All things considered, this chapter provides a foundation for comprehending how CIM is changing the construction industry in both developed and developing nations.

1.1 Background of Study

Construction is a dynamic and multifaceted process that involves assembling various components according to predefined design specifications to create physical structures in specific locations (Clough et al., 2015). It is an industry that spans across numerous disciplines and demands high levels of coordination among professionals such as architects, civil engineers, quantity surveyors, and project managers. As global economies continue to evolve, particularly with Asia's rapid industrialisation and the economic recovery of countries like the United States, the global construction sector is projected to grow at a pace surpassing that of the global GDP over the

© The Author(s), under exclusive license to Springer Nature Switzerland AG 2026
P. Adekunle et al., *Emerging Trends in Construction Information Management*,
https://doi.org/10.1007/978-3-032-09679-1_1

coming decade (Hook, 2015). This upward trajectory places increasing pressure on the construction industry to modernise and streamline its processes, particularly in how it manages information.

In emerging economies, the construction sector plays a critical role in economic development. It is a major source of employment, engaging approximately 5% of the total population, and contributes significantly to reducing unemployment. Construction projects typically involve a wide array of trades, tight timelines, and sometimes conflicting priorities. On-site collaboration is, therefore, not optional but essential for project success (Tabassi et al., 2012). Effective teamwork, characterised by open communication, shared responsibility, and mutual trust, is central to ensuring that construction sites operate efficiently and safely (Adekunle et al., 2022b). Team dynamics influence not only workplace safety but also project delivery outcomes and overall organisational performance (Huemann, 2010; Moe et al., 2010).

Beyond interpersonal collaboration, the complexity of modern construction projects has introduced new challenges in managing the vast and diverse types of data generated during project lifecycles. Unlike manufacturing, where processes are more repetitive and predictable, construction projects are often bespoke, leading to more complicated data flows (Pavan et al., 2014). This data includes technical drawings, material specifications, change orders, health and safety records, and compliance documentation, all of which require accurate and timely sharing among multiple stakeholders (Becker et al., 2011). However, traditional construction practices have often suffered from fragmented communication, inadequate stakeholder engagement, and inefficient information management systems (Palttala et al., 2012).

The challenge of poor information flow is particularly critical at different project stages, from design and procurement to construction and facility management, where any error or delay in data transmission can lead to rework, increased costs, and decreased productivity. Hendrickson (2008) highlights that the volume of information required even before breaking ground is vast, and this only increases as the project progresses. Errors related to miscommunication or data mismanagement negatively affect quality, time, and cost, thereby impacting labour productivity and client satisfaction (Adekunle et al., 2022a). These inefficiencies often lead to reduced labour productivity and diminished client satisfaction, particularly in complex, multi-stakeholder construction environments where timely and accurate information exchange is critical to success.

In response to these challenges, construction information management has undergone a significant transformation. Scholars and practitioners alike now advocate for adopting strategic approaches that align information technologies with core business processes. This includes implementing vision and policy frameworks, undertaking change management initiatives, reviewing new systems, and enhancing IT infrastructure (Kubjana et al., 2024). The goal is to foster continuous improvement and ensure long-term competitiveness through better knowledge sharing, real-time data access, and informed decision-making.

Technological advances have opened new avenues for enhancing construction information management (Spradley & Spradley, 2020). Building Information Modelling (BIM), artificial intelligence (AI), cloud computing, blockchain, and the

Internet of Things (IoT) have become increasingly relevant in reshaping how information is captured, shared, and utilised in construction projects. Globally, construction firms are leveraging these tools to develop innovative, low-cost business models, improve workflow automation, and reduce project risks (Farzad et al., 2019; Stewart, 2006). While these technologies are rapidly gaining traction in developed countries, their adoption in developing countries remains relatively slow, largely due to institutional barriers, skills shortages, and limited awareness of long-term benefits.

Given the construction sector's potential to drive economic growth, particularly in developing countries, there is an urgent need to explore and document emerging trends in construction information management. Understanding how digital tools and information strategies are transforming construction practices, both locally and globally, is vital for enhancing project delivery, improving industry competitiveness, and ensuring sustainable development. This study, therefore, aims to investigate the latest advancements, challenges, and opportunities associated with emerging trends in construction information management, with a focus on their implications for the construction industry in developing countries.

1.2 Problem Statement

The use of inappropriate or inaccurate information in construction projects can lead to serious consequences, including costly delays and, in some cases, complete project failure. Sanni-Anibire et al. (2019) assert that delays in construction often result in adverse outcomes such as legal disputes between contractors and clients, increased contract sums, loss of productivity and revenue, and disruptions in communication flow. A major factor contributing to these delays is ineffective management of construction information (Adekunle et al., 2024). One essential approach to addressing this issue is through Strategic Information Systems Planning (SISP), which helps organisations align their business strategies with information systems to achieve operational objectives.

Studies highlight numerous challenges in the management of construction information. These include poor communication channels (Winch et al., 2009), a lack of stakeholder engagement (Dadpour et al., 2019), and insufficient accountability structures (Galliers et al., 2014). Additionally, there is a lack of standardised methods for quantifying and forecasting project progress. This has hindered efforts to innovate and improve construction information systems, particularly at various project stages (Hajj et al., 2021). These shortcomings form significant barriers to achieving efficient and effective information management in construction.

However, recent technological advances, particularly those associated with the Fourth Industrial Revolution (4IR), offer promising solutions. The current generation's mindset, characterised by digital fluency and adaptability, presents a unique opportunity to address these long-standing issues (Cook, 2021). With technologies such as Building Information Modelling (BIM), Artificial Intelligence (AI), and the Internet of Things (IoT), construction processes can be significantly enhanced. The ability to share real-time data through integrated networks like 5G is transforming how construction data is gathered, processed, and applied (Dwivedi et al., 2021).

Although existing studies have addressed construction information management from various global perspectives, such as integration of design and construction knowledge (Eastman, 2003), application of knowledge-based systems (Chen & Kamara, 2011), and development of information modeling techniques (Chen & Kamara, 2008), there remains a gap in the literature concerning the context of developing economies. In particular, South Africa's construction industry faces unique challenges that demand context-specific research and solutions.

This study seeks to explore and assess the methods of managing construction information within developing economies. It will evaluate how these methods align with emerging technological tools and how they can be adapted to address local challenges. Furthermore, the study aims to identify strategies for improving communication, stakeholder collaboration, and data accountability within construction projects.

1.3 Aim and Objectives

From pre-contract to contract and then post-contract, the construction business requires a tremendous amount of information. This means that, aside from having creative and innovative ideas, there may be challenges in completing various types of building projects without the necessary information management system. There is a rising view that inadequate information systems are causing delays in project completion and an increase in abandoned projects in Sub-Saharan Africa. As a result, the current research intends to find construction information management methods. The study also looks into the problems and benefits of construction information management, as well as ways to improve construction information management.

1.4 Organization of the Book

1.4.1 Contribution and Value

This study will add to the body of knowledge by bringing together pertinent literature that supports construction information management methodologies, benefits, and problems. It will look into various technological gadgets as ways to improve construction data management. It will also help construction professionals recognize the value of a good and efficient construction information management system in the sector.

1.4.2 Summary

This research study was structured into six (6) separate chapters, which are titled as follows:

Chapter 1: Introduction

This chapter gave a framework for the whole study. It outlined the background of the subject of study, as well as the aim and significance. The introduction gave an overview of the problem statement, research methodology, limitations, research problems and research objectives.

Chapter 2: Information Management

This chapter reviewed published books, journals, dissertations, and articles, amongst others, on the same topic as or similar to those of other researchers and renowned scholars. The essence of this chapter was to provide a theoretical overview of information management. Hence, the evolution of information management was discussed, as well as information management areas, information management processes, the global concept of information management and the theoretical backgrounds of information management.

Chapter 3: Information Management in the Construction Industry

This chapter describes the methods of construction information management. Specific attention was given to the challenges and benefits of construction information management. The challenges and benefits highlighted are thoroughly discussed in this chapter.

Chapter 4: Methods of Improving Construction Information Management

This chapter presented a theoretical view of the various 4IR tools used in construction. An extensive explanation was provided on each of the mentioned tools. The potential of each tool to aid in the improvement of managing construction information was highlighted.

Chapter 5: Construction Information Management in Emerging Economies

This chapter introduced a hypothetical outline of the developing countries' construction industry and its role in fostering the economy. This chapter provides the readers with a comprehension of the idea of the state of the construction industry in selected developing countries. The chapter also discussed Building Information Modelling and Common Data Environment in the selected countries.

Chapter 6: Conclusions and Recommendations

This chapter enabled the researcher to conclude whether the research objectives had been achieved and whether the research questions had been answered. This chapter also made recommendations for future research.

References

Adekunle, P., Aigbavboa, C., Akinradewo, O., Oke, A., & Aghimien, D. (2022a). Construction information management: Benefits to the construction industry. *Sustainability, 14*(18), 11366. https://doi.org/10.3390/su141811366

Adekunle, P., Aigabvboa, C., Thwala, W., Akinradewo, O., & Oke, A. (2022b). Challenges confronting construction information management. *Frontiers in Built Environment, 8*, 1075674. https://doi.org/10.3389/fbuil.2022.1075674

Adekunle, P., Aigbavboa, C., Otasowie, K., & Ikuabe, M. (2024). Leveraging on technologies within quality management systems for enhanced project success in construction. In K. Arai (Ed.), *Proceedings of the Future Technologies Conference (FTC) 2024, Volume 4. FTC 2024.* Lecture Notes in Networks and Systems (Vol. 1157). Springer. https://doi.org/10.1007/978-3-031-73128-0_32

Akhavan, T. A., Mahyuddin, R., & Abu-Hassan, A. (2012). Effects of training and motivation practices on teamwork improvement and task efficiency: The case of construction firms. *International Journal of Project Management, 30*(2), 21–224.

Becker, T. C., Jaselskis, E. J., & McDermott, C. P. (2011). *Implications of construction industry trends on the educational requirements for future construction professionals* (pp. 1–12). Associated Schools of Construction.

Brede, M. N., Torgeir, D., & Tore, D. (2010). A teamwork model for understanding an agile team: A case study of a Scrum Project. *Information and Technology Software*, 480–491.

Chen, Y., & Kamara, J. (2008). The mechanisms of information communication on construction sites. *In Forum Ejournal, 8*(1), 1–32.

Chen, Y., & Kamara, J. M. (2011). A framework for using mobile computing for information management on construction sites. *Automation in Construction, 20*(7), 776–788.

Clough, R. H., Sears, G. A., Keoki, S. S., Segner, R. O., & Rounds, J. L. (2015). *Construction contracting: A practical guide to company management*. Wiley.

Cook, L. L. (2021). Insight into the millenial mind-set: Impact of 4IR and Society 5.0 on the real estate, construction and other industries. In *The ASOCSA 14th Built Environment Conference*, Florida (pp. 1–8).

Dwivedi, Y. K., Ismagilova, E., Sarker, P., Jeyaraj, A., Jadil, Y., & Hughes, L. (2023). A meta-analytic structural equation model for understanding social commerce adoption. *Information Systems Frontiers*, *25*(4), 1421–1437.

Eastman, C. M. (2003). *Integration of design and construction knowledge* (pp. 206–226). Routledge.

Farzad, J., Milad, Z., & Afshin, K. (2019). Life cycle environmental impact assessment to manage and optimize construction waste using Building Information Modeling (BIM). *International Journal of Construction Management*, 1–18. https://doi.org/10.1080/15623599.2019.1583850

Galliers, R. D., & Leidner, D. E. (2014). *Strategic information management: Challenges and strategies in managing information systems.* Butterworth-Heineman, Routeledge.

Hajj, C. E., Jawad, D., & Montes, G. M. (2021). Analysis of a construction innovative solution from the perspective of an information system theory. *Journal of construction engineering and management*, *147*(9), 03121003.

Hendrickson, C. (2008). *Project management for construction, fundamental concepts for owners, engineers, architects and builders*. Department of Civil and Environmental Engineering, Carnegie Mellon University.

Hook, J. (2015). *Global construction 2030—A global forecast for the construction industry to 2030* (pp. 1–13). Global Construction Perspectives and Oxford Economics, Broadwall House, London, United Kingdom.

Huemann, M. (2010). Considering human resource management when developing a project-oriented company: Case study of a telecommunication company. *International Journal of Project Management, 28*(4), 361–369.

Kubjana, L., Adekunle, P., Aigbavboa, C., & Akinradewo, O. (2024). Analyzing the impact of emotional intelligence on leadership in Construction 4.0. In K. Arai (Ed.), *Proceedings of the*

Future Technologies Conference (FTC) 2024, Volume 4. FTC 2024. Lecture Notes in Networks and Systems (Vol. 1157). Springer. https://doi.org/10.1007/978-3-031-73128-0_30

Mohammadhossein, D., Eghbal, S., & Ahad, N. (2019). Analysis of stakeholder concerns at different times of construction projects using social network analysis (SNA). *International Journal of Civil Engineering*, 1–15. ISSN 1735-0522. https://doi.org/10.1007/s40999-019-00450-1

Pauliina, P., Camillo, B., Ragnhild, L., & Marita, V. (2012). Communication gaps in disaster management: Perceptions by experts from governmental and non-governmental organizations. *Journal of Contingencies and Crisis Management, Volume 20*(1), 1–12.

Sanni-Anibire, M. O., Mahmoud, A. S., Hassanain, M. A., & Salami, B. A. (2019). A risk assessment approach for enhancing construction safety performance. *Safety Science, 121*, 15–29. https://doi.org/10.1016/j.ssci.2019.08.044

Spradley, E., & Spradley, R. T. (2020). *Simulating medical isolation: Communicatively managing patient and medical team safety* (pp. 45–48). Digital Commons.

Stewart, R. A. (2006). *IT enhanced project information management in construction: Pathways to improved performance and strategic competitiveness* (pp. 1–25). Griffith University.

Winch, S. A., Shumaker, R. W., Perkins, L., & De Vries, H. (2009). Captive and wild oraguntan (Pongo Sp.) survivorship: A cpmparison and the influence of management. *American Journal of Primatology: Official Journal of the American Society of Primatologists, 71*(8), 680–686.

Chapter 2
Overview of Information Management

Abstract This chapter examines the evolution of information management, tracing its historical development, the processes that shaped its emergence, and the current state of practice. It begins by exploring the early foundations of information management and how technological, organisational, and societal changes influenced its growth over time. The discussion highlights the transition from traditional manual approaches to advanced digital systems that enable more efficient data handling and decision-making in modern contexts. Furthermore, the chapter investigates the key areas of information management, including data collection, storage, analysis, dissemination, and security. Each area is explored to illustrate its role in supporting effective organisational operations and knowledge-driven strategies. Attention is also given to the processes that underpin information management, such as system design, governance, and integration, which ensure the effective use and protection of information assets. In addition, the chapter situates information management within a global context, recognising how globalisation and digital transformation have influenced practices across industries and regions. The theoretical background of information management is also reviewed, drawing from established frameworks and models that provide a foundation for understanding its principles. Ultimately, the chapter provides a comprehensive overview of information management as both a discipline and a practice.

2.1 Introduction

Information Management (IM) has evolved from rudimentary record-keeping practices, such as scribal archives and filing cabinets, into a sophisticated, strategic dimension of contemporary organisations, driven by technological advances and growing information complexity (Studocu, 2023; Teknita, 2024). Originally focused on organising physical documents and accounting ledgers, IM has expanded to encompass digital systems that support the full information lifecycle, from acquisition and storage to distribution and eventual disposal, ensuring that information remains accessible, secure, and actionable. The emergence of IM as a defined practice stems

© The Author(s), under exclusive license to Springer Nature Switzerland AG 2026
P. Adekunle et al., *Emerging Trends in Construction Information Management*,
https://doi.org/10.1007/978-3-032-09679-1_2

from two major drivers. First, economic and technological developments created vast quantities of information, precipitating what has been termed the "information crisis," wherein society needs systematic mechanisms to organise and leverage data effectively. Second, the interdisciplinary integration of management science, information theory, and system theory provided a conceptual foundation for IM to emerge as a structured field, combining both theoretical rigor and practical relevance. Within organisations, IM transcends operational tasks and plays a strategic role. It enables alignment between information systems and business objectives, informs decision-making, and enables organisations to adapt to complex environments. Theoretical frameworks underpinning IM draw from behavioural and organisational theory, particularly the concept of the "administrative man" and decision-making under constraints, and from economic theory, which contextualises IM within transaction and information-processing costs (Sophotree, 2023). This chapter explores the evolution, areas, processes, stages, and global perspectives of IM, before grounding these concepts in their theoretical underpinnings. It lays the groundwork for understanding how IM functions as both a disciplined practice and organisational capability with far-reaching strategic and operational implications.

2.2 Evolution of Information Management

Data management is a multi-layered discipline that is thought to look at the most recent approach to data management (Bouthillier & Shearer, 2002, p. 50). The competency and appropriateness of duties done by associations of the current information dependence is influenced by information management (Hicks, 2007, p. 233). As seen from the perspective of an association, ensuring that meaningful data is obtained and used to its full potential is a goal of data management (Willpower Information, 2005: 1). Data management is also based on an organization's movement pattern: obtaining data from at least one source, storing and exchanging data with end-clients, and finally removing data by discarding or storing it (Larson, 2005, p. 5). This data cycle differs across partners in an establishment, with the workforce in charge of value confirmation; data to be gained openness and viability; staff in charge of capacity security and disposal; and representatives who require it for decision-making (Treasury Board of Canada, 2006: 4).

Through the executives' approaches, these partners may be in a position to create, re-orchestrate, distribute, or remove data as indicated by the association's data. On a more visionary level, these elements can be thoroughly studied to add value to data by the element of how it is organised, conceptualised, and portrayed, as well as allowing data (thus worth) to float to end-clients through trade, sharing, and cooperation (Hicks, 2007, p. 234). Data management encompasses all of the executives' collective belief systems, which include the ability to develop, control, arrange, assess, coordinate, interact, and report data exercises, all of which are linked to meet the needs of the organization's roles and capacities that are data wards (Wang et al., 2006, p. 265).

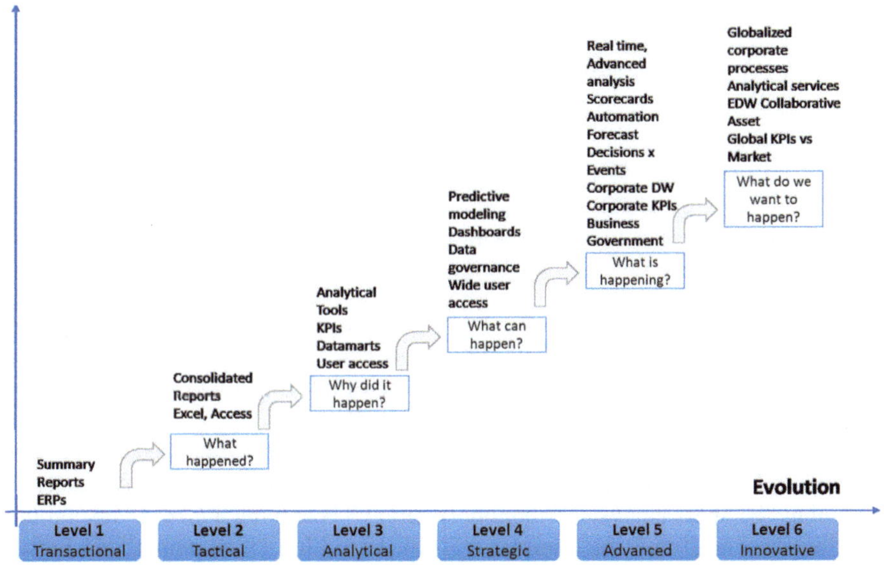

Fig. 2.1 Evolution of information systems at the analytical level. *Source* Guerrero and Sierra (2018: 2524)

The information that executives in educational organisations have used has centred on gathering important data for communication and recognising students and teachers (Guerrero & Sierra, 2018, p. 2523). The data have not been examined and are frequently backwards, confusing, redundant, and not reliable or complete (Guerrero & Sierra, 2018, p. 2523). This is the main flaw in current structures: there is no data organisation for organisational decision-making. For example, in the educational framework, administrators used data to acquire appropriate data for availability and limit the advancement of students and instructors (De Haan, 2006, p. 21). There was a necessity for a data update for the executives in this method. The executives' reaction to changes in business elements consists of data that continues to develop through time at the logical stage (Rashvanlouei et al., 2015, p. 70). Figure 2.1 illustrates in facet the evolution of information management at the critical level.

2.2.1 Emergent of Information Management

Information management was primarily viewed as aspects of what is now known as data management in the latter half of the twentieth century (Penn & Pennix, 2007: 25). Punched cards, magnetic tapes, and other media with a similar life-cycle in which information is created, distributed, backed up, maintained, and disposed of are used for record-keeping. Of this time, the current and future benefits of information technology were beginning to gain traction; a single chip could record a whole

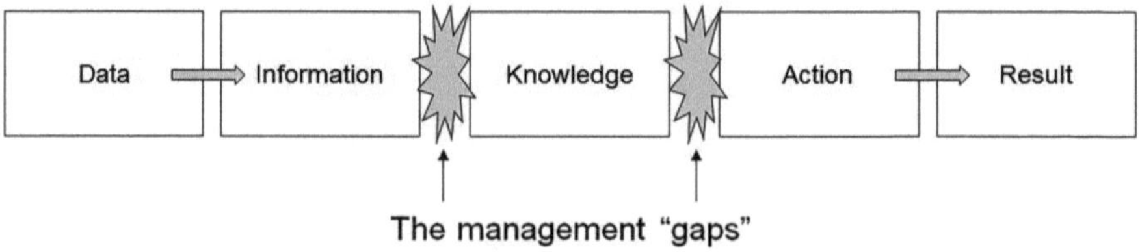

Fig. 2.2 Transformation of data. *Source* Ward and Peppard (2002: 207)

logbook or send messages all over the world via electronic mail, and important findings at the time (Gouseti, 2010: 352). Between 1980 and 1990, as information technology advanced and broadened the breadth of information systems, a new type of information management arose.

The lexicon of what was once known as "IT management" changed as a result of developing business enterprises such as British Crude-oil, with "systems analysts" becoming "business analysts," "monopoly supply" becoming a mix of "insourcing" and "outsourcing," and the complex IT task becoming "lean teams," paving the way for dexterity in the methods of gathering information for business advantage (Jeston, 2014: 25). The British broadened the breadth of senior management interest in information from value modeling to advanced business processes (O'Brien, 2010: 21). This technique is based on the efficiency with which information is managed, allowing the application of appropriate information systems to be outsourced after they have been operated on an IT infrastructure (Dibbern et al., 2004: 25). The transformation of data into various forms, as well as the role of management in achieving the intended result, are depicted in Fig. 2.2. It can be shown that management plays a significant role in the transformation of information into knowledge and from knowledge into action, which, if done incorrectly, can have an impact on the creation of the next stage (Ward & Peppard, 2002: 207).

2.2.2 *Positioning Information Management in the Bigger Picture*

Venkatraman (2005: 150) devised a basic methodology that simply outlines the management of knowledge, information, and data to explain the intermediate phase that lead to viewing information management strategically. He believes that data in the IT infrastructure industry should be defined in order to offer information. For the emergence of knowledge, it is critical to comprehend information within the information system. Knowledgeable managers make successful judgments, and the success of appropriate activities is dependent on effective decisions (Intezari & Pauleen, 2014: 395). It is sometimes referred to as "the DIKAR model: Data, Information, Knowledge, Act, and Result," and it was created to provide ideas about the layers involved in positioning technological innovation and organizational scheme.

It is seen as a pivotal moment in the transitional attitude toward information management (Bytheway, 2016: 28). It is critical that all digital businesses see information management as a tool for achieving significant results. Quality decision-making and information are critical to the organization's success (Bytheway, 2014: 27).

Figure 2.3 is a symbolic representation of the anatomy that depicts the five levels of IT-enabled organization transformation (Venkatraman, 2005: 146). Notably, these levels are not regarded as evolutionary steps because the actual mechanism does not conform to a single evolutionary phase's predetermined framework. The fundamental level for harnessing its capability within an industry is known as the introduction level. The term "localized exploitation" was chosen to describe how, in many cases, decisions about where to place remote systems (such as a client demand-entry system, toll-free client utility system, catalogue controls system, and intramural automated mail system) are redirected to the appropriate functional and operational executives (Weedmark, 2019: 1). As a result, there is little learning among the organization's executives, and such innovation is restricted.

Executives typically establish and place these systems in response to operational challenges or complications, such as a continuous, toll-free customer utility-support system to improve service (Donahue et al., 2000: 10). The next level is a logical extension of the previous one, demonstrating a more methodical attempt to influence IT proficiency throughout the entire business process. Interdependence processed through information (managing the interdependence of establishment title roles and responsibilities through distinct serviceable lines) and procedural interconnectivity (managing the interconnectivity and information sharing of the different systems and requisitions through a collective IT platform) (Agarwal & Brem, 2015: 7). Naturally, businesses pay more attention to the interconnection of information processes. Efforts towards systematic connectivity have been bolstered by significant advancements in connectivity capabilities over the last decade, such as increased availability of integrated technical resolutions and favorable cost-performance drift (Venkatraman,

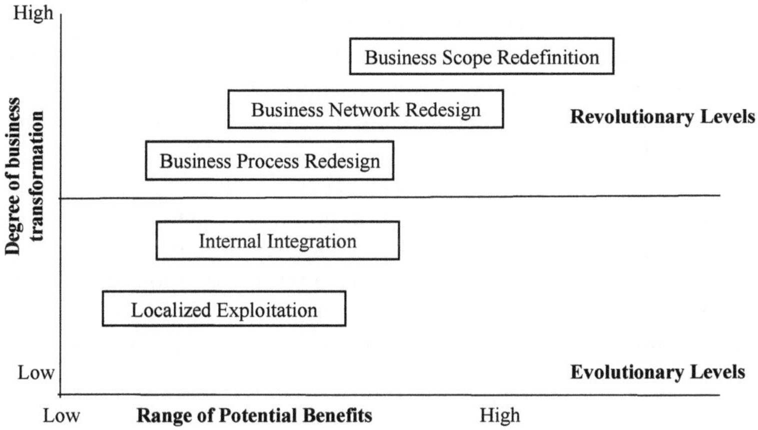

Fig. 2.3 Five levels of IT enabled business transformation. *Source* Venkatraman (2005: 146)

2005: 147). Every company must have a technical team (of varying informality) in charge of validating technological interconnection. Comparable and cross-functional groups only discuss the opposition of information procedural interdependence in very uncommon circumstances. The third stage, business process redesign, reflects a firm belief that the benefits of IT services will not be fully realized if they are forced on an existing business method, no matter how well integrated they are (Pellegrinelli et al., 2007: 44). The reason for this is because the current information system promotes a set of organizational philosophies that are in response to an industrial rebellion. Organizational perceptions such as centralism vs. decentralism, the scope of control, the line against staff, purposeful specialisation, authority accountability balance, organizational management processes, and sovereignty are all influenced by broad ideologies (Bose, 2012: 26).

Although these ideas are still valid, IT performance can substantially alter elements of these initial business process reform beliefs. Some of the shaping approaches may be considered to be ineffective. Experts and academics believe that an organization's new rationale should be built on current and developing IT competencies (Rojewski, 2002: 25). Business Network Redesign is the fourth phase. So far, the three steps discussed have focused on IT-assisted activity revolution within a single organization. These stages implied, either implicitly or explicitly, that the important organization's limit is fixed or predetermined. The circulation of information business actions among multiple enterprises is not modified in scenarios when there are interconnections with outside businesses such as providers, purchasers, and other mediators (Kano, 2018: 687). In contrast, this level denotes the reorganization of the state of interchange among many participants in a business network via the operational distribution of IT skills. The true impact of IT on any organization is determined not by reorganizing internal chores (efficiency gains), but by reorganizing the connections in long-distance business links to influence a larger range of proficiencies that will deliver more goods and services (Fallon, 2020: 2450). The quintal phase of the transformation is special in that any systematic endeavor to alter an establishment has ramifications for the establishment's business scope (Hanelt et al., 2020: 1163). The final phase is known as the redefining of the business scope. "What business are we in and why are we in it?" is a common starting point for tactic analysis (Marr & Creelman, 2011: 23). "What role (if any) does IT play in influencing business scope and the logic of business interactions inside the extended business network?" asks the quintal phase of transformation. Newer models such as shared ventures, coalitions and partnerships, and cybernetic business links with obvious importance directed at a more elastic and liquidated corporate scope are replacing older models such as economies of scale (within the pecking order), product-line extension led through straight-up incorporation, and mergers procurements that steered to increased prominence on straight-up incorporation. At this point, the emphasis is on the specific enabling nature of information technology inside this approach.

The transition of business links (level four) from contract processing to knowledge grids has resulted in continuous inferences about the rationality of operations and the resulting redistribution of proceeds and revenue (fringe) streams in a given market

(Larsen & Klishewski, 2004: 5). The reason for this is that some responsibilities may be eliminated (such as tedious quality control stages, billing charges, and formulating supply slips), others may be reorganized optimally through organizational constraints (combined design or collective production), and still others may be stretched (value-added functions embedded in IT functions) (Venkatraman, 2005: 146).

2.3 Information Management Areas

Information management is a broad term that encompasses policies and procedures for centrally handling and disseminating data among various individuals, organizations, and/or information schemes throughout the information life cycle (Sarafeimidis & Smithson, 2003: 255). Information management is primarily a systematised information system perception endeavour in which an organization develops, owns, and maintains a collection of data (Moussipoulous et al., 2010: 379). The data could be in the form of substantial data (such as papers, pamphlets, and records) or computerized data (Simon, 2017: 3). The use of an organization's authority to influence and regulate its information virtues is known as information management (Luenendonk, 2017: 2). Information management is typically accomplished using purpose-built information management schemes as well as supporting business procedures and policies (Sage & Cuppan, 2001: 333). IM also focuses on how information is disseminated and conveyed to multiple recipients, such as characters and various electronic devices, such as a company's website, computers, internet servers, software, and/or mobile devices (Geel et al., 2013: 108). The following are the various aspects of information management:

Information Technology

The speed with which innovation changes and the pressure to always have the most recent innovation items might compromise the stability of the foundation that supports frameworks, thereby assisting business processes and attaining benefits (Larson & Chang, 2016: 8). It's critical to deal with the "supply side" and recognize that innovation is always becoming a requirement.

Information System

While most data frameworks were developed internally, in the long run, it has become possible to obtain the majority of the product plots that an establishment requires from the product abridgment industry (Diamond & Khemani, 2005: 8). By and large, there is a sense of impending major addition from the application of new frameworks and ideas that relate to the strategic goals of organizations.

Business Processes and Business Information

Data structures are useful to business systems in terms of fostering them, and they pass on information to the business in the form of business data (COA, 2009: 2). Because it isn't widely used, business process management is still considered as a relatively new theory. As a rule, it's been extreme; managing corporate data is really challenging.

Business Benefit

It is vital not only to be brutally honest about potential outcomes, but also to confirm the active surveillance and valuation of benefit distribution (Markus & Tanis, 2000: 176). There has been a lot of mindfulness in business direct administration since the approach and universalization of the balanced scorecard, but not a lot of genuine effort has been made to link business lead of executives to the compensations of data innovation financing and the beginning of inventive data frameworks until this thousand years (Cooper et al., 2017: 1019).

Business Strategy

Regardless of whether to deal with poor execution or boost variety and seriousness, many foundations' strategy must be influenced by data innovation and data planning possibilities (Kavanagh & Thite, 2017: 60). Tactical analysis tools like the significance chain and acute achievement factor assessment rely on paying close attention to the data that is (or could be) controlled (Bai & Sarkis, 2013: 286).

2.4 Information Management Processes

The information management (IM) process entails a series of activities involving the collection and management of data from at least one source, as well as the dissemination of that data to at least one audience (Pickard et al., 2021: 1). Data is an organized arrangement of images that record or express a message in its most specialized form. The core idea is that data is a collection of facts that have been organized in such a way that they have more value than the actual realities (Techopedia, 2021: 1). The frameworks engineer is both the source and the recipient of data items; as a result, the frameworks engineer has a significant stake in the success of the turn of events and the effective use of IM interaction and frameworks (Pickard et al., 2021: 1). Designing IM is the first step in the IM cycle. The IM technique or plan is the result of this process. The following step is to do IM. The construction, populating, and maintaining at least one data store, as well as the generation and dissemination of data managements reports, are the results of these methods (I-scoop, 2021: 1).

2.4.1 Plan Information Management

When structuring data, four key criteria should be considered: scope, constraints, control/security, and life cycle (Mosley, 2009: 3). The extension is the key component, with a few sub-factors to consider. The sub-factors include the recognizable proof of the data to be monitored, the length of time the data should have been stored, and whether a framework information word reference is required to label data for ease of search and recovery (O'Neal, 2017: 2). Physical, electronic, or a combination of media will be used to oversee the data. When putting up information archives, how was a work in progress (WIP) and officially provided data previously viewed? (Sacks et al., 2010: 641).

The imperatives are the next factor to evaluate, and they, too, include a few sub-factors (Mosley, 2009: 4). The following sub-factors are identified: the degree of arrangement control that must be applied to the data, the administrative prerequisites relating to the management of data for the project (including commodity control requirements), inspect client requirements or arrangements relating to the management of project data, identification of any industry guidelines relating to the management of task data, and recognisable proof of a project (Boutilier et al., 2012: 232).

Control and security are the third factor to consider while organizing data (Mosley, 2009: 4). Clients, partners, providers, and administrative specialists are among the sub-factors that have access to the data: those working on the project, other members of the organization/venture, clients, partners, providers, and administrative specialists (Stair & Reynolds, 2021: 12). Another sub-factor is the requirements to protect the data from unauthorized access, which includes licensed innovation and freedoms that must be considered—for example, if data from providers is to be stored and the possibility of a provider accessing data having a place with another project contender provider (Barnett, 2016: 13). Other sub-factors to consider include the information archive or storehouses to be used, the volume of data to be stored when choosing vaults, and the speed of access and search when selecting stores (Uddin et al., 2020: 3019). Apart from the aforementioned sub-factors, the document design took into account the electronic data to be stored, the defined requirements for data storage to be legal, and defined requirements to ensure that data is discarded effectively when it is not generally needed to be stored or when it is no longer significant (for example, defining the audit time frame for each snippet of data) (Reardon, 2013: 308).

The Life Cycle is the final variable to consider when arranging data. The life span of the electronic data to be saved (for example, accessibility of nonpartisan document configurations, formation of product duplicates, or holding of pre-owned data), the requirement for disaster recovery (in the event that a server holding electronic data is annihilated, are there back-up duplicates of the data, consistently have to demonstrate that data recovery is faultless) are some of the sub-factors (Pickard, 2021: 4).

2.4.2 *Perform Information Management*

The combination of ICT speculation with business process reengineering increases the positive yield consequences (Devaraj & Kohli, 2003: 45). According to a survey of company leaders, the crucial integration of ICT ventures with business systems results in increased business value from ICT investments (Tallon et al., 2000: 161). The monetary exhibition of the 'effective clients' of ICT is higher than that of the 'less fruitful clientele.' They got at this conclusion by comparing the monetary exhibitions of 100 organizations identified by Computerworld/IDG as 'effective clients' of ICT to the monetary exhibitions of 100 comparative organizations designated as 'less fruitful clients' of ICT (Stratopoulos & Dehning, 2000: 44). Taking an asset-based approach to the firm entails not only an interest in ICT framework, but also the development of one-of-a-kind ICT abilities that lead to improved firm performance (Bharadwaj, 2000: 173). The combination of decentralization practices (assigning greater decision power, forming independent groups, and increasing specialist obligations) and ICT has a lopsidedly enormously good effect on firm market value (Brynjolfsson et al., 2000: 12).

The main driver of the effect of ICT on monetary and non-monetary execution, according to a longitudinal report in clinics, isn't curiosity in innovation, but the actual use of innovation (Devaraj & Kohli, 2003: 281). His findings show that ICT is a key empowering agent of representative inclusion and all-out-quality administration, and that their blend with ICT contributes decisively to the exhibition of firms, while exploring the effect of ICT and three arrangements of authoritative work practices: worker association, all-out quality administration, and reengineering (Ramirez, 2003: 32). ICT capital, new hierarchical patterns, and human resources all contribute significantly to work efficiency, according to data from Swiss companies. It also demonstrated the complementarity of ICT capital and human resources: the combined use of ICT and human resources results in increased work efficiency beyond the individual effects of these two variables (Arvanitis, 2003: 34). The "ICT and Economic Growth" study from the Organisation for Economic Co-operation and Development (OECD) provides data on ICT dispersion as well as substantiation of ICT's positive impact on macroeconomic, sectoral, and firm levels in OECD countries. It argues that ICT is necessary for a broader range of modifications that aid in better implementation. However, the consequences of ICT rely on correlative hypotheses, such as in correct capacities and hierarchical modifications, such as new systems, new business procedures, and new authoritative structures' (OECD, 2003: 91). Brynjolfsson and Hitt (2000: 42) arrive at resolutions that are comparable. Both contextual and econometric studies point to authoritative supplements, such as new business procedures, new abilities, and new hierarchical and industrial structures, as a major driver of data innovation commitment. These related ventures, and the resulting resources, could be as much as a substantial degree larger than the interest in PC innovation itself.

Pickard et al. (2021: 11) went even further, highlighting elements to consider as part of the data management process:

- The data's legitimacy (accessibility to the data management procedure/plan and a list of data to be monitored);
- The work process for data audit and endorsement is defined as moving data from "work in progress" to "delivered."
- The correct arrangement for the executive's needs when it comes to data;
- Determining whether or not the data is being baselined;

- The appropriate "labels" are applied to the data to account for simple hunt and recovery;
- The data is subjected to the appropriate access regulations;
- Before capacity, determining if the data is being translated into an unbiased document design;
- An audit date is specified for checking the data's authenticity after it has been processed;
- The work procedure for surveying and removing unwanted, invalid, or mysterious data (as defined in association/venture strategy, project strategy, security, or protected innovation requirements) is being defined;
- Examining whether the data is credible and whether the reinforcement recovery framework is being used;
- Checking to see if the data is correct and if the reinforcement recovery system is being used;
- Examining whether the IM framework meets defined execution requirements, such as speed of access, accessibility, and accessibility.

2.4.3 Information Management Cycle

The concept of data management as a pattern that is between connected data exercises to be expected, planned, and constructed provides a cycle-based perspective to the firm's board of directors that supports the more traditional perspectives on data management as data innovation or data asset (Scherer & Schapke, 2011, p. 586). Starting with the identification of data wants and continuing through data procurement, association and capacity, goods and administrations, appropriation, and closing the cycle with data use, the interaction model of data management should cover the entire data esteem chain (Sellitto et al., 2007, p. 75). In most cases, data management solutions do not include needs identification and data utilization (Hicks, 2007: 237). Despite the fact that needs investigation is one of the most overlooked phases of data management, the nature of the data that the client receives is heavily influenced by how successfully the requirements have been communicated (Petter et al., 2008: 240). Also, data use is important because understanding how data is used (or not used) to make decisions, solve problems, or decipher situations is critical for continuous development of the executive's other data-driven activities (Alavi & Leidner, 2001: 121). Figure 2.4, in addition to this discussion, depicts the data management operations as discrete boxes arranged in a straight line. In reality, the cycles are not easily segmented, the exercises cross across, and their boundaries are fluid.

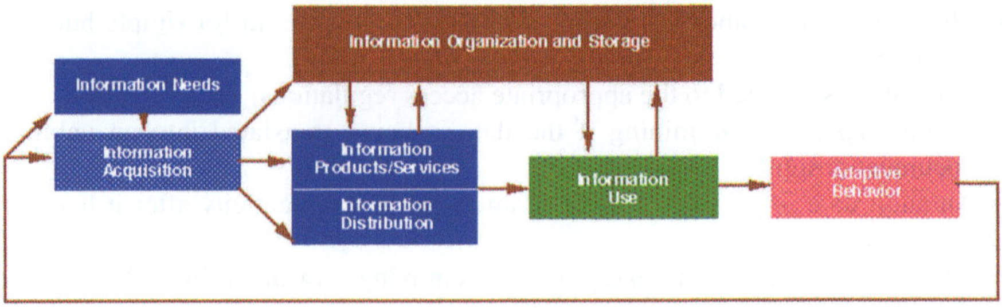

Fig. 2.4 Information management cycle. *Source* Choo (2012: 242)

2.5 Stages in Information Management

Establishments should focus more on managing sensitive information, such as inferred knowledge, judgment, and instinctual capacities, to achieve viable data management (Busch et al., 2002: 2). The information management implementation process has five stages (Naybour, 2015: 1). This strategy is recommended by the majority of consulting firms and vendors.

Information Planning

The first step is to create a data model, which outlines how data will be managed as part of the task (Naybour, 2015: 1). It will usually lay out the duties for data management, the systems to be followed, and the finer points, such as how to number records and drawings (Bhattacherjee & Sanford, 2006: 811). This is one of the task the management plan's tactics and techniques.

Information Capture

The next stage is data collection, which is the project manager's responsibility (Naybour,2015, p. 1). When selecting the data to be captured (for example, only important data to be acquired), the project manager must exercise caution (Voinov et al., 2016, p. 201). This is significant since arranging, exploring, and dissecting any approaching material takes time and effort. For example, acquiring non-essential information for leaving utilities outside the project region (Hertzum & Pejtersen, 2000: 763).

Information Storage

Data storage is the third stage. Data can be found in a variety of formats, including written versions, messaging, music, video, and so on. Electronic capacity is commonly used these days since it allows for the storage of large amounts of data and allows for the maintenance of backups for stored data (Baker et al., 2006: 225). Put away data should be properly recorded in a non-editable format with adaption control (Niezen et al., 2016: 180). Access to project colleagues' vital information was restricted; for example, many as-constructed designs for the work site's current utilities were gathered in printed forms (Zou et al., 2017: 55). These are inspected and saved on the project server, with duplicates forwarded to the offer point connection for use by different colleagues (Khumalo and Mearns, 2019: 2).

Information Reporting

The task administrator is responsible for converting vital information into data, which can be summarized as brevity in achieving objectives, opportuneness, and precision (Gunasekaran et al., 2004: 337). When that is completed, data can be shared among the task group for use or communicated to the support and partners in a way that they can understand (Sutterfield et al., 2006: 29). For example, a project manager may announce assistance with crucial data associated with a pressing issue.

Information Archiving

Data archiving is the final stage. Data should be kept for a specific period of time that should be agreed upon with the client (Montes et al., 2013: 2029). The project manager must also decide on how the information will be documented (for example, printed versions or delicate duplicates), how frequently the information will be recovered (especially for printed version records), and the time it will take to recover the reports (Bressan & Canazza, 2013: 2). For example, in the organization's Dubai store, project as-assembled drawings and reports are documented in printed copies (Shafiq, 2021: 130). It will take at least two working days to recover any of the information that has been filed (Naybour, 2015: 1).

2.6 Global Concept of Information Management

As a result of technological advancements and the widespread distribution of data, many organizations are experiencing the negative impacts of data over-burdening and are required to use data management to control the data turbulence in the computerized world (Dias, 2001: 1). Most of the time, this data is stored in a disorganized manner in personal computer equipment, spread out in data sets, making access to essential information difficult and reducing representatives' usefulness in their day-to-day activities (Morabito, 2013: 23). As a result, many cutting-edge projects lack a global perspective on their own data and information (Wielki, 2017: 266). The corporate gateway coordinates unstructured information to organized information from functional data sets using metadata1 and eXtensible Markup Language2 (XML), allowing access to corporate data via a customized interface available over the inward hypertext organization; the Intranet (Roth et al., 2002: 565). The key business data the executives undertake in the coming decade is the corporate doorway (Landquist & Stenmark, 2007: 119). The corporate entryway, which evolved from information distribution centers, now extends its application to the Intranet and serves as a single point of access to all data and information assets in a company (Chan & Chung, 2002: 114).

2.6.1 Strategic Information Management

As data frameworks that provide that ability became formalised and robotised, capabilities were seriously tested at many levels, according to the Carnegie Mellon School. An organization's capacity to deal with data is at the center of authoritative and administrative skill, and an association's methodology should be intended to further develop data handling ability (Saldanha et al., 2017: 270). Associations were thought to be expected to learn and modify in never-before-seen ways. Scholastics began to organize and disseminate definitive works on the critical management of data and data frameworks (Hendrickson et al., 2013, p. 232). Simultaneously, the management of business processes and information executives' ideas, although being a big part of the early hopeful contemplation of business process upgrading, have since been damaged in the data management writing (Wendy, 2004, p. 240). The understanding of the data climate, defined as the total of persons, organisations, and frameworks that gather, process, disseminate, or follow up on data, is considered critical in the field of important exams (Roche et al., 2013, p. 25). This atmosphere is made up of three interconnected elements that are constantly communicating with people, organisations, and frameworks (Cantarello et al., 2012, p. 32). Physical, instructional, and intellectual aspects are all present.

A Contemporary Portfolio Model for Information

In a single book, Bytheway (2014: 27) has gathered and organized basic instruments and methods for data collection across the board. A portfolio model that considers the flood of interest in outer wellsprings of data and the need to organize unorganized data outside to make it useful is at the heart of his data management perspective (Fig. 2.5). This portfolio approach addresses both internal and external obtaining concerns, as well as the management of data, which might be structured or unstructured.

Bytheway went on to say that an information portfolio like this one demonstrates how information may be obtained and organized in four stages:

Stage 1: Using public data; recognize and use all around organized external plans of reference information, such as post codes, climatic data, GPS positioning data, and trip itineraries, as exemplified in the individualized computer press (Ashbrook & Stamer, 2003: 285).

Stage 2: Tagging the commotion on the internet; use existing plans such as post codes and GPS coordinates, or, more commonly, by adding labels, or develop a suitable philosophy that provides structure (Shirky, 2005: 1). Shirky presents an overview of these two approaches.

Stage 3: Sifting and investigating: in a larger world, the summed-up ontologies, which are still a work in progress, extend to many elements and many relationships among them, providing the resources to elicit significance from massive amounts of data. When the structure of organized data sets resembles a higher level data model, a metaphysics, or an element connection model, it works best (Noy & McGuiness, 2001: 23).

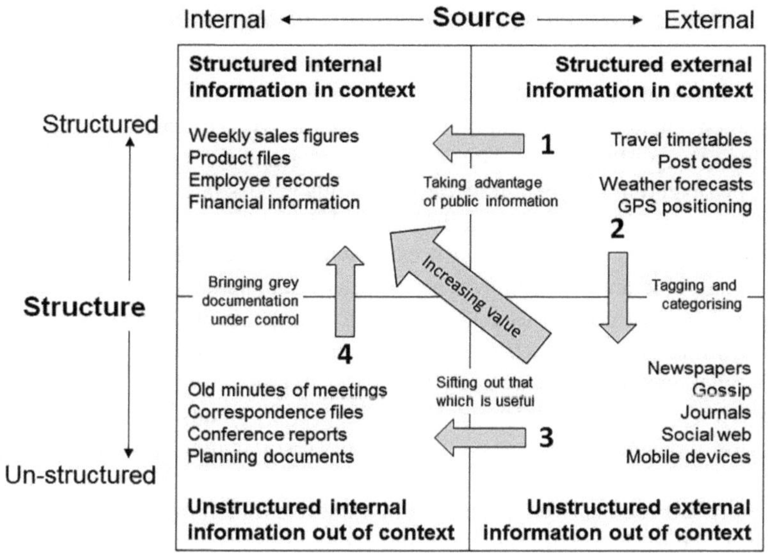

Fig. 2.5 Information portfolio. *Source* Bytheway (2014: 29)

Stage 4: Data structuring and archiving: given the vast amount of data available from sources such as the social web and the smaller-than-expected telemetry systems used in close-to-home health care, improved techniques to document and then fish data for significant data are required (Haggan et al., 2009: 631). Guides diminish strategies, which begin with useful writing computer programs, are a later method of evoking data from massive authentic datasets that are attracting the attention of ordinary organizations with exceptionally large data assets to work with, but they require advanced multi-processor assets (Chu et al., 2007: 281).

There are many different ways to look at a firm, and the data from the perspective of the management is simply one of them. Keep in mind that distinct areas of business movement will also contribute to the system. It's not only good data management that propels a company forward. Corporate management, human resource executives, product development, and marketing will all play important roles in the future. We should not regard a single movement space as the exclusive source of vital achievement. Corporate management, human resource executives, product development, and advertising, on the other hand, all depend on effective data management. On the broad premise stated here, the ability to manage data well can be said to be transcendent in the final examination.

2.6.2 Operationalising Information Management

At the functional level, establishments are regularly confronted with several data management challenges and dilemmas, particularly when hierarchical change occurs (Wiredu, 2012, p. 190). The unusual nature of new framework designs and the board's lack of involvement in recent data developments necessitate a level of authoritative change among CEOs that is notoriously difficult to explain (Voldman & Alblas, 2012, p. 271). There will be a persistent breakdown in data processes due to an overall hierarchical reluctance to empower new types of data management architecture. For example, a lack of critical assets, a failure to recognise new types of data and the new ways that use them, a lack of support from senior management leading to a lack of vital vision, and unexpected political manoeuvring sabotaging the entire organisation's activities (Morton & Hu, 2008, p. 395). In any event, the execution of new forms of data management should normally result in functional benefits (Pereira, 2009, p. 375).

Five critical regions were identified in a data handling viewpoint on association strategy to build data handling limit and reduce the necessity for data processing (Galbraith, 2012, p. 3), as shown below:

- Creating slack assets to reduce the heap on the general pecking order of assets and to lessen data handling associated with over-burden;
- Creating independent assignments with defined limits and that can achieve legitimate conclusion and with all of the current assets needed to play out the undertaking;
- Recognize parallel relations that cut across utilitarian units to move choice capacity to the interaction rather than dividing it within the order;
- Invest in vertical data frameworks that route data streams for a specific assignment (or set of undertakings) according to the applied business rationale.

The concept of sidelong relations leads to a hierarchical structure that differs from the basic progression, the lattice association (Ramezan, 2011: 90). This brings together an organization's vertical (multiple tiered) and even (item or venture) perspectives on work that is visible to the rest of the globe. The establishment of a network association is one administration response to the continuous smoothness of outside interest, avoiding various and incorrect responses to roundabout inquiries that will typically be managed independently (Pici, 2017: 253).

2.7 Theoretical Background

A data management system provides a strategy heading, as well as the necessary standards and best practices, to ensure that data management fulfills hierarchical requirements (NetMonkey, 2009: 1; Andersen & Mandrich, 2007: 1). Many studies appear to agree on the need for a procedural configuration for any organization to deal with its data, as many of them provide a method for dealing with business data (Maes, 2000: 1; Lake, 2009: 4). Preparation (Blumenthal, 2009: 1), quality and accessibility (Infokits, 2007: 1), data innovation (Lake, 2009: 1), overseeing creation, usage, and recovery (COA, 2009: 1), security the executives (Infokits, 2007: 1), and overseeing data dispose of are among the elements mentioned above (Andersen & Mandrich, 2007: 1). The force of any of the pieces varies depending on the size of the relationship and the conditions (Jennings, 2007: 1). According to Jennings (2007: 1), full authoritative responsibility for monitoring data as a venture resource necessitates the formation of large business agreements and the creation of a strong environment shared by the association to oversee data. In any case, figuring out what it implies and where to start might be difficult. On similar lines, there are data management aids such as advancements, procedures, and improvement instruments; but, as valuable as these are, they must be docile to hierarchical data methodology and strategy (Bruckner, 2016, p. 276).

2.7.1 Behavioural and Organisational Theory

It is widely agreed that good data is critical to the successful operation of organizations, and while there is no universally approved management theory, conduct and authoritative speculations can aid (Mackenzie et al., 2005: 715). One important aspect of data handling and navigation is an individual's ability to manage data and make decisions under limits imposed by the situation (Elfenbein & Knott, 2015: 960). Models can be an individual's age, the complexity of a situation, or a lack of essential quality in the data that is currently available, all of which is exacerbated by the rapid development of innovation and the new types of framework that it enables, especially as the social web emerges as a peculiarity that business cannot ignore (Bengler et al., 2014: 10). However, data management must be considered as useful frameworks, with an unquestionable degree of data handling and an enormous necessity for decision making at various levels, for a long time before there was any widespread recognition of the importance of data management in establishments (Raghu & Vinze, 2007: 1066). In light of their argumentation concerning the intellectual furthest reaches of soundness, they presented "authoritative man" as an alternative to the "financial man" model, which is backed by traditional hypothesis (Klaes & Sent, 2003: 16). They also proposed the concept of satisficing, which entails sifting through the available possibilities until an adequate edge is reached, which is another money-related idea (Gonzalez-Valdes & Ortuzar, 2018: 79).

2.7.2 Economic Theory

Regardless of the social and authoritative perspective, there are a variety of challenges that stem from financial and natural factors (Colombo & Grilli, 2013: 399). There's the cost of acquiring and analyzing the facts needed to make a decision, as well as the time and effort involved (Metzger, 2007: 2080). Data cycles can come at a significant cost in terms of currency. Set up authoritative principles and procedures, in particular, can prevent you from making the best decision, resulting in less-than-ideal outcomes (Naidoo et al., 2011: 1145). This is a problem that has been identified as a significant concern with administrative organizations that are losing important change economies due to entrenched mentalities (Lawal, 2007: 6).

The "data handling" and "decentralization of motivators" streams of financial considerations of hierarchical plan building provide useful bits of knowledge about the topic viable here (Colombo et al., 2015: 495). While the data handling stream ignores organizational difficulties, it emphasizes that everyone has a limited ability to obtain, store, process, and deliver data (Joseph & Gaba, 2019: 65). As a result, the expansion of a center administration level for a two-layer firm overseen by a proprietor results in an increase in data handling capacities, resulting in two benefits (Colombo & Grilli, 2013: 401).

First and foremost, data overburden issues that impede dynamic action inside the company are addressed (Slotegeraaf & Atuahene-Gima, 2011: 103). In a two-layer progressive system, centralizing all decisions with the proprietor administrator reduces the number of decisions made per unit of time as compared to a situation where a few decisions are decentralized to center directors and made simultaneously (Laurea, 2016: 7; Miyajima, 2012: 5). Second, a three-tiered chain of command protects the proprietor supervisor as she attempts not to waste her valuable time on low-value additional options that can be delegated to center administrators (Harris & Raviv, 2002: 856). It featured the observational benefit of delegating a center supervisor that comes from the lowering of the regular range of control, while anticipating that the destinations of center directors are aligned with those of the head (Ng & Szeto, 2015: 8). More recently, there has been a focus on improving the quality of choices made possible by establishing a central administration level. Workers frequently have personal information that offers them a data advantage over the owner/manager in certain situations (Wong & Aspinwall, 2004: 47). This specific premise should be co-situated with dynamic power in order to make valid decisions in terms of data management.

2.8 Lessons Learnt

This chapter's literature review begins with a brief history of the emergence of information management, which stretches back several decades. Information management's goal, as seen from the standpoint of an organization, is to ensure that important information is earned and properly utilized. The information management that is commonly used in educational organizations is focused on gathering relevant data for communication and monitoring students and educators. Naturally, businesses pay more attention to the interconnection of information processes. Efforts at systematic connectivity have been bolstered by significant advancements in connectivity skills over the last decade, such as increased availability of integrated technical resolutions and favorable cost-performance drift. Data management is a comprehensive concept that entails coordinating ways and measures for halfway managing and disseminating data among many people, foundations, and data plans throughout the data life cycle. Data management is largely a drive toward ordered data framework discernment, in which a foundation develops, owns, and manages a variety of data. The data could be in the form of paper, pamphlets, or records, or it could be in the form of electronic information assets. Data management takes advantage of an association's position of power and control over its data ideals. Data management is widely recognized as a result of rationale produced inside executives' objectives and sponsorship of business procedures and norms.

Data, in its most specialized sense, is an organized set of images that record or deliver a message, according to the writing investigation. The core idea is that data is a collection of facts that have been organized in such a way that they have more value than the actual realities. The frameworks engineer is both the source and recipient of data items; as a result, the frameworks engineer plays a critical role in the successful execution of the IM cycle and IM frameworks. Designing IM is the first step in the IM cycle, and then performing IM is the next step. Executives and establishments should focus more on controlling sensitive information, such as inferred knowledge, judgment, and natural capacities, in order to provide convincing data. The data management execution method is divided into five stages: data arranging, data collection, data storage, data detailing, and data filing. Many organisations are experiencing the negative impacts of data overburden because of technological advancements and the wide dispersion of data, and they must employ data across the board to manage this data chaos in the computerised world. Frequently, this data is haphazardly stored in personal computer equipment and dispersed across databases, making access to essential information difficult and hurting workers' efficiency in their day-to-day tasks. Theoretical foundation also suggests that full authoritative responsibility for overseeing data as an endeavor resource necessitates the development of big business approaches and the creation of a shared help climate by the association to oversee data, despite the fact that figuring out what that means and how to get everything started can be extremely difficult.

References

Agarwal, R., & Brem, A. (2015). Strategic management of technology and innovation: A review and implications for future research. *Technological Forecasting and Social Change, 98*, 1–6. https://doi.org/10.1016/j.techfore.2015.06.001

Alavi, M., & Leidner, E. D. (2001). Review: Knowledge management and knowledge management system: Conceptual foundations and reserach issues. *MIS Quaterly, 25*(1), 107–136.

Andersen, J., & Mandrich, L. (2007). *Information life cycle* (pp. 1–2). Microsoft Coorporation.

Arvanitis, S. (2003). The role of information management in knowledge economies. *International Journal of Information Management, 23*(1), 3–13. https://doi.org/10.1016/S0268-4012(02)00067-8

Ashbrook, D., & Stamer, T. (2003). Using GPS to learn significant locations and predict movement across multiple users. *Personal and Ubiquitous Computing, 7*(5), 275–286.

Bai, C., & Sarkis, J. (2013). Green information technologies and systems: A framework and literature review. *International Journal of Production Economics, 147*, 516–529. https://doi.org/10.1016/j.ijpe.2013.03.028

Baker, M., Shah, M., Rosenthal, D. S. H., Roussopoulos, M., Maniatis, P., Guili, T. J., & Bungale, P. (2006). *A fresh look at the reliability of long-term digital storage* (pp. 221–234).

Barnett, J. M. (2016). Three Quasi-fallacies in the conventional understanding of intellectual property. *Journal of Law, Economics & Policy, 12*(1), 1–46.

Bengler, K., Dietmayer, K., Farber, B., Maurer, M., Stiller, C., & Winner, H. (2014). Three decades of driver assistance systems: Review and future perspectives. *IEEE Intelligent Transportation Systems Magazine, 6*(4), 6–22.

Bharadwaj, A. (2000). A resource based perspective on information technology capability and firm performance: An empirical investigation. *MIS Quaterly, 24*(1), 169–196.

Bhattacherjee, A., & Sanford C. (2006). Influence process for information technology acceptance: An elaboration likelihoof model. *MIS Quaterly, 30*(4), 805–825.

Blumenthal, D. (2009). Health information technology in the era of health reform. *New England Journal of Medicine, 363*(11), 1080–1081. https://doi.org/10.1056/NEJMp1001057

Bose, R. (2012). Business data communications and networking. Business Information Systems Series. *Routledge.*

Bouthillier, F., & Shearer, K. (2002). Understanding knowledge management and information management: The need for an empirical perspective. *Information Research, 8*(1), paper 141. http://informationr.net/ir/8-1/paper141.html

Boutilier, R. G., Black, L., & Thompson, I. (2012). *From metaphor to management tool: How the social licenece operate can stabilise the socio-political environment for business* (pp. 227–237).

Bressan, F., & Canazza, S. (2013). A systematic approach to the preservation of audio documents: methodology and software tools. *Journal of Electrical and Computer Engineering, 4*(2), 1–22.

Brückner, S. (2016). Digital construction: Transforming the industry. *Journal of Construction Engineering and Management, 142*(6), 04016019. https://doi.org/10.1061/(ASCE)CO.1943-7862.0001146

Brynjolfsson, E., Lorin, H., & Shinkyu, Y. (2000). *Intangible assets: How the interaction of computers and organizational structure affects stock market valuations* (p. 3). MIT Working Paper.

Bytheway, A. (2014). Digital information governance in organisations. *Journal of Information Governance, 1*(2), 45–60.

Bytheway, A. (2016). Organizational information governance frameworks. *Journal of Information Governance, 3*(1), 15–32

Busch, P., Richards, D., & Dampney, C. K. (2002). Tacit knowledge characteristics: A research methodology. *Association for Information Systems Proceedings, 77*, 1–16.

Cantarello, S., Martini, A., & Nosella, A. (2012). Knowledge management and innovation: The role of organizational context and national culture. *Journal of Knowledge Management,16*(4), 581–599. https://doi.org/10.1108/13673271211246140

Chan, M. F. S., & Chung, W. W. C. (2002). A framework to develop an enterprise information portal for contract manufacturing. *International Journal of Production Economics, 75*, 113–126.

Chu, C., Kim, S. K., Lin, Y., Yu, Y. Y., Bradski, G., Ng, A. Y., & Olukoyun, K. (2007). Map-reduce fror machine learning on multicore. *Advances in Neural Information Processing Systems, 19*, p. 281.

Choo, C. W. (2012). Designing effective organizational knowledge strategies: A framework for understanding information management and knowledge management. Oxford UniversityPress.

COA-Paper. (2009). *Information life-cycle management.* [Online] Available at: http://www.jerichoforum.com/informationmanagementprimer

Colombo, M. G., & Grilli, L. (2013). The creation of a middle-management Levy by entrepreneurial ventures: Testing economic theories of organisational design. *Journal of Economics & Management Strategy, 22*(2), 390–422.
Colombo, M. G., Rossi-Lamastra, C., & Matassini, B. (2015). The organisation design of high-tech entrepreneurial ventures. *Foundations and Trends in Entrepreneurship, 11*(6), 427–523.
Cooper, R., Grey, S., Raymond, G., & Walker, P. (2017). Project risk management guidelines: Managing risks in projects. *International Journal of Project Management, 35*(5), 840–853. https://doi.org/10.1016/j.ijproman.2016.10.007
De Haan, J. (2006). Knowledge management and organisational culture in the construction sector. *Construction Management and Economics, 24*(10), 1103–1113. https://doi.org/10.1080/01446190600950482
Devaraj, S., & Kohli, R. (2003). Information technology payoff in the health care industry: A longitudinal study. *Journal of Management Information Systems, 16*(4), 283–322.
Diamond, J., & Khemani, R. (2005). Governance and public sector information management. Public *Administration Review, 65*(S1), 117–126. https://doi.org/10.1111/j.1540-6210.2005.00460.x
Dias, L. P. (2001). Information management in construction projects: A strategic approach. *International Journal of Project Management, 19*(3), 155–162. https://doi.org/10.1016/S0263-7863(00)00012-3
Dibbern, J., Goles, T., Hirschheim, R., & Jayatilaka, B. (2004). Information system outsourcing: A survey and analysis of the literature. *Database for Advances in Information Systems, 35*(4), 1–25.
Donahue, R., Zeckhauser, R., & Johnson, R. (2000). Managing public information: Lessons from government agencies. *Public Administration Review, 60*(3), 217–226. https://doi.org/10.1111/0033-3352.00028
Elfenbein, W. D., & Knott, M. A. (2015). Time to exit: Rational, behavioural, and organisational delays. *Strategic Management Journal, 36*, pp. 957–975.
Fallon, R. (2020). Digital transformation in construction: A review and future research directions. *Automation in Construction, 112*, 103085. https://doi.org/10.1016/j.autcon.2020.103085
Galbraith, J. R. (2012). Organizing to deliver solutions. *Organizational Dynamics, 41*(2), 82–91. https://doi.org/10.1016/j.orgdyn.2012.02.003
Geel, K., Winch, G., & Langford, D. (2013). Knowledge transfer in construction project networks. *Construction Management and Economics, 31*(1), 34–45. https://doi.org/10.1080/01446193.2012.717737
Gonzalez-Valdes, F., & Ottuzar, J. D. D. (2018). The stochastic satisficing model: A bounded rationality discrete choice model. *Journal of Choice Modelling, 27*, 74–87.
Gouseti, A. (2010). Information systems and the management of construction projects. *International Journal of Project Organisation and Management, 2*(1), 27–44. https://doi.org/10.1504/IJPOM.2010.030879
Guerrero, C., & Sierra, M. (2018). Digital data practices in construction project management. *Journal of Engineering, Design and Technology, 16*(2), 256–273. https://doi.org/10.1108/JEDT-06-2017-0065
Gunasekaran, A., Patel, C., & McGaughey, R. E. (2004). A framework for supply chain performance measurement. *International Journal of Production Economics, 87*, 333–347.
Haggan, N., Jackson, G. D., & Laeroix, P. (2009). Salmon and eulachon in ecosystem space and time: a plea for wider collaboration and data integration. In *American Fisheries Society Symposium* (Vol. 69, pp. 619–636).
Hanelt, A., Piccinini, E., Gregory, R. W., & Kolbe, L. M. (2020). Digital transformation and related research: A bibliometric analysis and a research agenda. *Journal of Business Research, 118*, 200–212. https://doi.org/10.1016/j.jbusres.2020.06.032
Harris, M., & Raviv, A. (2002). Organisation design. *Management Science*, 852–865. https://doi.org/10.1287/mnsc.48.7.852.281

Hendrickson, C., Au, T., & Tommelein, I. (2013). Managing construction information flows: Tools and techniques. *Journal of Construction Engineering and Management, 139*(10), 04013019. https://doi.org/10.1061/(ASCE)CO.1943-7862.0000786

Hertzum, M., & Pejtersen, A. M. (2000). The information-seeking practices of engineers: Searching for documents as well as for people. *Information Processing and Management, 36*, 761–778.

Hicks, B. J. (2007). Lean information management: Understanding and eliminating waste. *International Journal of Information Management, 27*(4), 233–249.

I-Scoop. (2021). Digital transformation definition and meaning. https://www.i-scoop.eu/digital-transformation/

Infokits. (2007). Digital information management in construction projects. *Infokits Publishing*. http://www.infokits.org

Jeston, J., & Nelis, J. (2014). Business process management: Practical guidelines to successful implementations (3rd ed.). Routledge.

Jennings, M. (2007). *Developing a roadmap for an entreprice information management program.* [Online] Available at: http://www.eimi-archives/volume-1-issue-1-march-2007-edition/enterprise-information-management-primer

Joseph, J., & Gaba, V. (2019). Organisational structure, information processing, and decision making: A retrospective and roadmap for research. *Academy of Management Annals, Forthcoming,* 1–91.

Kano, N. (2018). Kano model: Customer satisfaction and product development. *International Journal of Quality & Reliability Management, 35*(9), 2105–2124. https://doi.org/10.1108/IJQRM-01-2018-0012

Kavanagh, M. J., & Thite, M. (2017). Digital strategies and HR in construction organizations. *Human Resource Management Review, 27*(3), 483–495. https://doi.org/10.1016/j.hrmr.2016.10.004

Khumalo, S., & Mearns, M. (2019). Sharepoint as enabler for collaboration and efficient project knowledge sharing. *South African Journal of Information Management, 21*(1), 1–9.

Klaes, M., & Sent, M. (2005). A conceptual history of the emergence of bounded rationality. *History of Political Economy, 37*(1), 27–59.

Landquist, F., & Stenmark, D. (2007). Challenges and pitfalls in portal information management. In *Encyclopedia of portal technologies and applications* (pp. 118–122). IGI Global.

Larson, E. W. (2005). Project management and information systems integration in construction. *Journal of Construction Engineering and Management, 131*(12), 1280–1287. https://doi.org/10.1061/(ASCE)0733-9364(2005)131:12(1280)

Larsen, E. W., & Klishewski, J. (2004). Knowledge flows in construction project networks. *Construction Management and Economics, 22*(1), 41–52. https://doi.org/10.1080/0144619032000160567

Larson, E. W., & Chang, C. Y. (2016). Project management and information systems in construction: Integrating BIM and supply chain. *Construction Innovation, 16*(1), 4–22. https://doi.org/10.1108/CI-04-2015-0012

Laurea, T. D. (2016). *The reform of Halian mutual banks: The implications on their operations* (pp. 1–112). University of Padua.

Lawal, G. (2007). Corruption and development in Africa: challenges for political and economic change. *Humanity and Social Science Journal, 2*(1), 1–7.

Luenendonk, M. (2017). Information management frameworks: A practical overview. https://www.luenendonk.de/information-management-frameworks/

Maes, R. (2000). Information management strategies for construction firms. *International Journal of Construction Information Technology, 8*(2), 77–94.

Marr, W. (2011). Key performance indicators for information systems. Business Expert Press.

Marr, B., & Creelman, J. (2011). Big data in practice: How 45 successful companies used big data analytics to deliver extraordinary results. Wiley.

Markus, M. L., & Tanis, C. (2000). The enterprise systems experience — from adoption to success. In R. W. Zmud (Ed.), Framing the domains of IT management: Projecting the future through the past (pp. 173–207). Pinnaflex Educational Resources.

Mackenzie, S. B., Podsakoff, P. M., & Jarvis, B. C. (2005). The problem of measurement model misspecification in behavioral and organisational research and some recommended solutions. *Journal of Applied Psychology, 90*(4), 710–730.

Metzger, M. J. (2007). Making sense of credibility on the web: Models for evaluating online information and recommendations for future research. *Journal of the American Society for Information Science and Technology, 58*(13), 2078–2091.

Miyajima, H. (2012). *Institution change and its economic consequence in Japan: The bright and dark sides of hybridization* (pp. 1–45). The Research Institute of Economy, Trade and Industry.

Montes, J., Sanchez, A., Memishi, B., Perez, S. M., & Antoniu, G. (2013). GMonE: A complete approach to cloud monitoring. *Future Generation Computer System, 29*(8), 26–35.

Morabito, V. (2013). An information management approach emphasising on the "I" in IT. In *Business Technology Organisation* (pp. 23–44). Springer.

Morton, N. A., & Hu, Q. (2008). Implications of the fit between organisational structure and ERP: A structural contingency theory perspective. *International Journal of Information Management, 28*(5), 391–402.

Mosley, M. (2009). *The DAMA guide to the data management body if knowledge (DAMA-DMBOK Guide).* Tecnics Publications (ISBN9781935504023).

Moussipoulous, N., Papadopoulos, A., & Koulouriotis, D. (2010). ICT adoption in construction SMEs. *European Journal of Information Systems, 19*(3), 198–207. https://doi.org/10.1057/ejis.2010.9

Naybour, R. (2015). Digital transformation and information flows in construction. *Construction Innovation, 15*(2), 98–112. https://doi.org/10.1108/CI-04-2014-0023

Naidoo, R., Shankar, A., & Veer, E. (2011). The consumerist turn in higher education: Policy aspirations and outcomes. *Journal of Marketing Management, 27*(11–12), 1142–1162. https://doi.org/10.1080/0267257X.2011.609135

NetMonkey. (2009). *The importance of information management.* [Online] Available at: www.netmonkeys.co.uk/whitepapers/information-managemet.pdf

Ng, S., & Szeto, S. E. (2015). Preparing school leaders: The professional development needs of newly appointed principals. *Educational Management Administration and Leadership*, pp. 1–18. https://doi.org/10.1177/1741143214564766

Niezen, G., Vrijhoef, R., & Koskela, L. (2016). Information flows and collaboration in construction projects. *Engineering, Construction and Architectural Management, 23*(5), 575–596. https://doi.org/10.1108/ECAM-07-2015-0107

Noy, N. F., & McGuiness D. L. (2001). *Ontology Development 101: A guide to creating your first ontology.* Stanford Knowledge systems laboratory technical report KSL-01-05 and Stanford medical informatics technical report SMI-2001-0880.

O'Brien, J., & Marakas, G. (2010). Introduction to information systems (15th ed.). McGraw-Hill.

O'Neal, K. (2017). *What is difference between information management and data management* (pp. 1–2). First San Francisco Partners.

OECD. (2003). *Information and communications technologies; ICT and economic growth: Evidence from OECD countries, industries and firms.* Organisation for Economic Co-operation and Development.

Pellegrinelli, S., Partington, D., Hemingway, C., Mohdzain, Z., Shah, M., & Sydow, J. (2007). The role of projects in organizational change. *International Journal of Project Management, 25*(3), 209–217. https://doi.org/10.1016/j.ijproman.2006.11.009

Penn, I. A., & Pennix, G. B. (2017). *Record management handbook.* Routledge. https://doi.org/10.1108/JICES-03-2021-0036

Pereira, J. V. (2009). The new supply chain's frontier: Information management. *International Journal of Information Management, 29*, 372–379.

Petter, S., Delone, W., & McLean, E. (2008). Measuring Information System success: Models, dimensions, measures and interrelationships. *European Journal of Information System, 17*, 236–263.

Pici, R. (2017). Digital workflows and construction site management. *Construction Management and Economics, 35*(7), 427–439. https://doi.org/10.1080/01446193.2017.1301234

Pickard, A. (2021). Information governance and records management in the digital age. *Records Management Journal, 31*(4), 256–273. https://doi.org/10.1108/RMJ-04-2021-0010

Raghu, T. S., & Vinze, A. (2007). A business process context for knowledge management. *Decision Support System, 43*, 1062–1079.

Ramezan, M. (2011). Intellectual capital and organisational organic structure in knowledge society: How are these concepts related? *International Journal of Information Management, 31*(1), 85–95.

Ramirez, R. (2003). *The influence of information technology and organisational improvement efforts on the performance of firms*. Doctoral Dissertation, University of California, California.

Rashvanlouei, A., Hosseini, M. R., & Stewart, R. A. (2015). BIM adoption and information management in construction. *Automation in Construction, 57*, 155–165. https://doi.org/10.1016/j.autcon.2015.05.010

Reardon, K. K. (2013). Knowledge management in construction: Lessons learned. *Journal of Knowledge Management, 17*(3), 345–363. https://doi.org/10.1108/JKM-01-2013-0011

Roche, S., Propeck-Zimmermann, E., & Mericskay, B. (2013). Geo Web and Crisis Management issues and perspectives of Volunteered geographic information. *Geo Journal, 78*(1), 21–40.

Rojewski, J. (2002). Managing construction information systems: An integrated approach. *Construction Information Technology Review, 4*(2), 23–39.

Rojewski, J., & Smith, D. (2002). Information management for construction projects. *Construction Management Journal, 8*(1), 12–29.

Roth, M. A., Wolfson, D. C., Kleewein, J. C., & Nelin, C. J. (2002). Information integration: A new generation of information technology. *IBM Systems Journal, 41*(4), 563–577.

Sacks, R., Koskela, L., Dave, B. A., & Owen, R. (2010). Interaction of lean and building information modeling in construction. *Journal of Construction Engineering and Management,136*(9), 968–980. https://doi.org/10.1061/(ASCE)CO.1943-7862.0000199

Sacks, R., & Harel, O. (2017). BIM and information management in construction. *Automation in Construction, 81*, 37–51. https://doi.org/10.1016/j.autcon.2017.03.009

Sage, A., & Cuppan, C. (2001). Framework for managing complex projects. *International Journal of Project Management, 19*(1), 1–8. https://doi.org/10.1016/S0263-7863(99)00046-6

Sarafeimidis, D., & Smithson, S. (2003). Information strategy and project performance in construction. *International Journal of Project Management, 21*(6), 815–826. https://doi.org/10.1016/S0263-7863(02)00092-9

Saldanha, T. J. V., Mithas, S., & Krishnan, M. S. (2017). Leveraging customer involvement for fueling innovation: The role of relational and analytical informatiion processing capabilities. *MIS Quarterly, 41*(1), 267–286.

Scherer, R. J., & Schapke, S. E. (2011). A distributed multi-model based Management Information System for simulation and decision making on construction projects. *Advanced Engineering Informatics, 25*(4), 582–599.

Sellitto, C., Buress, S., & Hawking, P. (2007). Information quality attributes associated with RFID-derived benefits in the retail supply chain. *International Journal of Retail and Distribution Management, 35*(1), 69–87.

Shafiq, M. T. (2021). Client-driven level 2 BIM implementation: A case study from the UAE. *The Open Construction and Building Technology Journal, 15*, 129–139.

Shirky, C. (2005). Social software and the politics of groups. First Monday, *10*(6). https://www.firstmonday.org/article/view/1230/1142

Simon, H. A. (2017). The sciences of the artificial (3rd ed.). MIT Press.

Slotegraaf, R. J., & Atuahene-Gima, K. (2011). Product development team stability and new product advantage: The role of decision-making processes. *Journal of Marketing, 75*(1), 96–108.

Sophotree. (2023). Managing construction knowledge and data. https://www.sophotree.com

Stair, R., & Reynolds, G. (2021). Part 1—Information systems in business society. *Principles of Information System* (pp. 3–28). Cengage Learning.

Stratopoulos, T., & Dehning, B. (2000). Does successful investment in information technology solve the productivity paradox? *Information & Management, 38*(2), 103–117.

Studocu. (2023). Studocu World University Ranking 2023. *Studocu.* https://www.studocu.id/id/world-university-ranking/2023.

Sutterfield, S. J., Friday-Stroud, S. S., & Shivers-Blackwell, S. L. (2006). A case study of project and stakeholder management failures: Lesson learned. *Project Management Institute, 37*(5), 26–35.

Tallon, P., Ktaemer, K., & Gurbaxani, V. (2000). Executive perception of the business value of information technology: A process oriented approach. *Journal of Management Information System, 16*(4), 145–173.

Techopedia. (2021). Digital transformation definition. https://www.techopedia.com/definition/33377/digital-transformation

Teknita. (2024). Digital information management insights for construction projects. https://www.teknita.com

Treasury Board Secretariat (Treasury Board of Canada). (2006). Information Management in the Government of Canada. https://www.tbssct.canada.ca/im-gi/resources/tv-lv/tv-lvpr-eng.asp. tbs-sct.canada.ca+1

Uddin, M. A., Stranieri, A., Gondal, I., & Balasubramanian, V. (2020). Rapid health data repository allocation using predictive machine learning. *Health Informatics Journal, 26*(4), 3009–3036.

Veldman, J., & Alex, A. (2012). Managing design variety, process variety and engineering change: A case study of two capital good firms. *Research in Engineering Design, 23*(4), 269–290.

Venkatraman, N. (2005). IT-enabled business transformation: From automation to business scope redefinition. *Sloan Management Review, 45*(4), 73–87.

Voinov, A., Kolagani, N., McCall, M. K., Glymn, P. D., & Ramu, P. (2016). Modelling with stakeholders-next generation. *Environmental Modelling & Software, 77*, 196–220.

Wang, X., & Huang, J. (2006). The relationships between key stakeholders' project performance and project success: Perceptions of Chinese construction supervising engineers. *International Journal of Project Management, 24*, 253–260.

Ward, J., & Peppard, J. (2002). Strategic planning for information systems (3rd ed.). John Wiley & Sons

Weedmark, K. (2019). Records management and information governance for construction firms. *Records Management Journal, 29*(2), 101–119. https://doi.org/10.1108/RMJ-06-2019-0024

Wendy. (2004). Information systems management in construction. Wendy Publishing.

Willcock, S., & Henderson, K. (2014). Information management maturity in construction firms. *International Journal of Construction Education and Research, 10*(4), 264–279. https://doi.org/10.1080/15578771.2014.929036

Willpower Information. (2005). Organizational information strategies. *Willpower Information.* http://www.willpowerinformation.org

Wielki, J. (2017). The impact of the internet of things concept development on changes in the operations of modern enterprises. *Polish Journal of Management Studies, 15*(1), 262–275.

Wiredu, G. O. (2012). Information sysem innovation in public organisations: An institutional perspective. *Information Technology and People*, 188–206.

Wong, Y. K., & Aspinwall, E. (2004). Characterising Knowledge management in the small business environment. *Journal of Knowledge Management, 8*(3), pp. 44–61. ISSN 1367-3270. https://doi.org/10.1108/13673270410541033

Zou, P. X. W., Lun, P., Cipolla, D., & Mohamed, S. (2017). Cloud-based safety information and communication system in infrastructure construction. *Safety Science, 98*, 50–69.

Chapter 3
Information Management in the Construction Industry

Abstract This chapter explores the critical role of Information Management (IM) within the construction industry, mapping its lifecycle and examining diverse management methods. It investigates traditional systems, Computer-Integrated Construction, Continuous Acquisition and Life-Cycle Support, and Project Management Information Systems, highlighting their evolution and integration. Major challenges, including issues related to design data, cost, operational inefficiencies, and workforce limitations, are identified through academic and industry research, particularly noting legacy systems, inadequate technological infrastructure, and leadership shortfalls as key impediments. Conversely, the chapter demonstrates how effective information management delivers tangible benefits: enhanced communication, streamlined contact points, operational efficiency, improved cost and schedule control, and amplified market insights. The analysis underscores the transformation of IM from a primarily administrative process to a strategic capability that supports project success, organisational performance, and competitive advantage. The chapter concludes by distilling practical lessons for industry stakeholders seeking to bridge the gap between theoretical best practices and real-world application.

3.1 Introduction

The construction industry, renowned for its complexity and multitude of stakeholders, generates and relies on vast streams of information, from design documents and schedules to cost estimates and quality reports, that must be managed effectively throughout the project lifecycle (Adekunle et al., 2022). Proper Information Management (IM) is not merely an administrative necessity; it serves as the backbone for delivering construction projects efficiently, safely, and economically. Historically, IM began with paper-based systems and manual control of drawings and contracts, creating silos and making version control nearly impossible. Today, construction projects commonly suffer from fragmented information, poor coordination, unnecessary rework, and cost overruns, symptoms of weak IM structures (Tambwe et al., 2025). A telling example is the persistent challenge of preserving

© The Author(s), under exclusive license to Springer Nature Switzerland AG 2026

P. Adekunle et al., *Emerging Trends in Construction Information Management*,
https://doi.org/10.1007/978-3-032-09679-1_3

information integrity across phases, from design to handover, where documents are lost, misinterpreted, or converted across incompatible formats, resulting in reduced productivity and elevated failure costs.

IM processes are emerging to counter these inefficiencies through structured methodologies. Methods such as Computer-Integrated Construction, automation of document workflows, integration of lifecycle data, and the deployment of Project Management Information Systems (PMIS) represent a move toward more cohesive, data-driven management. These systems support critical functions like document control, schedule and cost tracking, and automated reporting, enabling stakeholders to make timely, informed decisions. Nevertheless, multiple challenges remain entrenched. A survey by Adekunle et al. (2022) highlights internal constraints like sustained reliance on legacy systems, insufficient investment in technology and infrastructure, weak leadership engagement, and poorly structured policies as significant barriers to effective IM. These reinforce the fragmented nature of the industry and undermine IM efforts.

However, when effectively implemented, IM yields substantial benefits. It promotes improved communication and coordination among project teams, serves as a reliable point of contact and information reference, and enhances operational efficiency. There is clear evidence that structured IM contributes to better schedule and cost management, elevated organisational performance, greater market competitiveness, and bolstered employee motivation, ultimately enhancing the quality of services delivered.

3.2 Information Cycle in the Construction Industry

Members of development projects are typically divided into five categories: customer, general project worker, subcontractors, advisors, providers, and related organizations (Zhiliang & Liang, 2004). In an ideal world, a variety of authoritative management hierarchies may be created by combining these individuals in various ways (Vallas & Christin, 2018). For example, there is a cycle in which the overall contractor forms a total agreement with the client, followed by the contractor signing contracts with the subcontractors. In another cycle, the customer may sign an agreement with the worker for hire for the main components of the project, and then sign the excess pieces of the task with the sub-project workers. During the development phase of this cycle, it is common for subcontractors hired to focus on planning and execution in accordance with that of the contractor (Bolumole).

Data management exercises are done on administrative protests and can be arranged in a variety of methods during development (Isikdag, 2012). The administrative components can be ordered as time, cost, quality, security, agreements, material and machines, human resources, and so forth (Lee & Yu, 2012). Furthermore, the undertaking components of the administration exercises can be divided into two categories in terms of data management understanding. One is the record-keeping task, which necessitates the entry of data, such as establishing cost-cutting measures for executives (Lin & Su, 2013). This kind of assignment typically includes a handling sequence, such as submission, review, and approval. The other is a duty that does not require data contribution, such as checking the outcome obtained by processing the management structure. Under specific circumstances, an unrecorded undertaking could become a record-involved errand. When a surprise result is observed, for example, the perceptual result may need to be documented.

In the construction industry, the data management cycle has become a major concern for professionals. In response to the rising number of reported occurrences, researchers and inventors have created a slew of new technologies to facilitate information flow while maintaining professionalism and anonymity (Wender & Hubler, 2009). In construction risk-incident management, the capacity to quickly locate key information is critical. Due of the abundance of construction information, construction administrators must filter data and ensure that it is correct and succinct (Leite et al., 2011).

3.3 Methods of Construction Information Management

Projects in the construction business are becoming increasingly large and complex, with new development advancements, tactics, and other such things emerging at a rapid pace (Lee et al., 2018: 1). Furthermore, developers are increasingly varying their expectations, while all related laws and norms are rapidly changing (Jung et al., 2013: 90). Because of such advancements in development innovation and the growing complexity of the development industry, skilled administration of the many data generated by development initiatives is becoming increasingly important (Choi, 2011: 35). Since the development stage is when the structure begins to take shape based on the plans, an effective task necessitates far-reaching management of massive amounts of data in order for the development stage to succeed (Jung & Joo, 2013: 80). Because of the importance of data, executives in the development industry employ a variety of information technology procedures, tests, and techniques, such as CIC (Computer Integrated Construction), CALS (Continuous Acquisition and Life-cycle Support), and PMIS (Project Management Information System), to deal with the data generated during development (Jung & Kim, 2016: 91). However, looking at these investigations makes it difficult to use the data because the data created in the development industry is gathered as a single snippet of data, such as e-reports (Jung & Woo, 2004: 619). They exclusively focus on information perception and, as a result, are insufficient.

3.3.1 Traditional Information Management System

This traditional data management strategy works brilliantly for the expert researcher who contributes somewhat independently to their workgroup's research endeavors in material science, engineering science, science, or, in any case, design (Kuhne & Liehr, 2009, p. 18). Moving this data to associates or replacements, on the other hand, is wasteful, given that each researcher sets up simple and advanced data vaults with an individualised, efficient request. The ensuing weakness grows in proportion to the number of workgroup members and staff changes, which is quite large, especially in academic domains. As a result, numerous researchers and PhD students have been confronted with the problem that older estimating data cannot be deciphered because the total history of the data cannot be recreated, e.g., which tests were estimated, which exploratory boundaries were applied, or which properties, such as the atomic weight and polydispersity of polymer tests, could not be determined (Hartnett & Rew, 2018, p. 3). Obviously, such issues can arise as a result of a variety of factors, the most common of which are fragmented or incorrect paperwork. As a result, basic solutions are ineffective. However, the documentation is generally adequate, but the overall setting is lost due to the large number of steps to be reported separately and the complicated connections between individual simple and electronic records, such as conventions in lab diaries, essential information documents on hard plates, and so on (Olsen et al., 2012: 8). As a result, the beginning of explicit data can no longer be totally rebuilt or backtracked, rendering them useless for true logical work (Tempini, 2017: 204). This is a waste of time and effort because the estimates must be rehashed in the majority of cases. Make use of a method that prevents such frictional misfortunes (Herb, 2008: 36). However, this method should preferably not interfere with the researchers' current working strategy, but rather be based on standard structures, as experience has shown that alterations needing a significant amount of change are not recognized by mainstream researchers (Hodgson, 2001: 45).

Traditional data management and refinement procedures include a number of consecutive and occasionally covering stages (Choo, 2002: 156). The cycles are determined by why a particular entertainer produces data into the framework (for example, data framework, data set), after which the data is evaluated and combined with other data and information, then sent to customers and used in navigation (Vuori & Okkonen, 2012: 120). The cycles are typically cyclic; utilizing the data generated during the process usually elicits new data requirements, resuming the cycle. Like serious knowledge (CI) processes, business-arranged data the board procedures typically follow a similar structure, such as a pattern of back-to-back stages (Vitt et al., 2002: 228). Such cycles are based on specific predeterminism, such as predetermined data or information requirements, which determine the outcome. Two examples of common data management interaction models are shown in Fig. 3.1. Because of the shared data, refining the cycle isn't as clearly defined, and it wouldn't be possible to show it in the same way each time because entertainers may change or their feedback may differ. McKelvie et al., (2007: 410) emphasize the possibility of rapid transformation of new practices, particularly those based on impromptu tasks.

The interaction can be described as dark and omnipresent: black due to a high degree of variability, universal due to the fact that only a few (sub) processes are involved inside a refining task. The feasibility of such a common method of doing things can be clarified through synergistic argumentation (Brown & Renshaw, 2000:

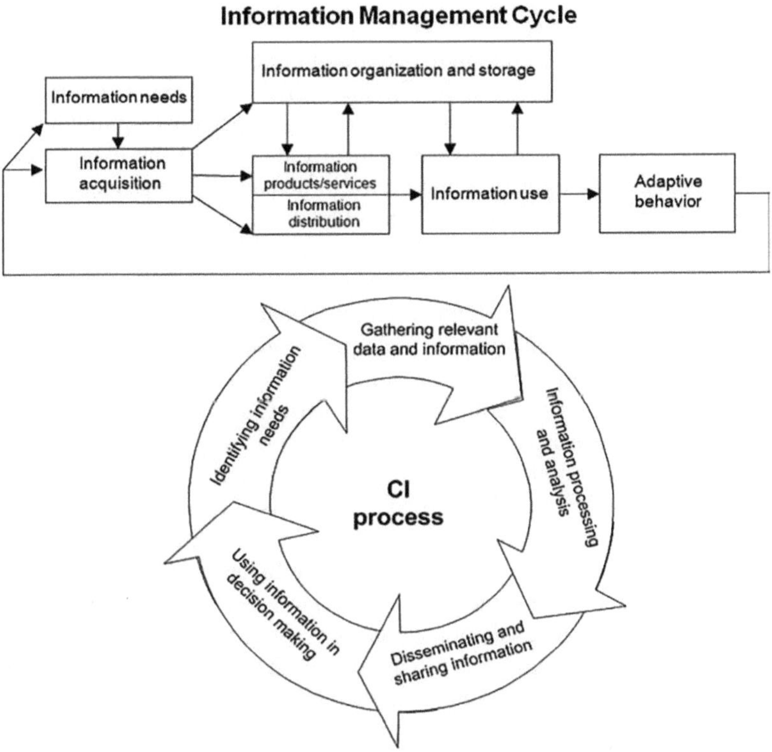

Fig. 3.1 Examples of traditional information management processes featuring the Information Management Cycle and a generic CI process. *Source* Vuori and Okkonen (2012: 120)

55) or sense-making in business systems, for example (Kurtz & Snowden, 2003: 465). This can be considered as prod usage when looking at web-based media in a corporate scenario, such as concurrent content creation and use (Grudin, 2006: 64). As shown in Fig. 3.1, the data refinement process enabled by online media apps can be represented as numerous whirls signifying nonstop and simultaneous prod usage and interaction (Vuori & Okkonen, 2012: 120).

3.3.2 Computer Integrated Construction

There has been a long history of research into using data innovation in the development field, which is commonly referred to as Computer integrated construction (CIC) (Eastmen et al., 2007: 11). The potential for quality and usability advances through industrial technology support has been perceived in several industries. All things considered, the development region has frequently been delayed to exploit, in contrast to other enterprises (Davis & Songer, 2002: 288). When compared to other planned item endeavors such as airplanes and automobiles, the development space is claimed to have lagged behind in the adoption of coordinated frameworks for the planning and production of its outputs (Rezgui et al., 2000: 138). To combat some of these challenges, a lot of research has been done to give personal computer support to working on the correspondence between plan information and plan expectation (Kazi et al., 2001: 131). For example, a number of initiatives led by academic scientists, standardization agencies, and industry consortiums have developed information/item models aimed at facilitating information exchange between programming applications (Katranuschkov et al., 2001: 123). Various initiatives have developed word references, thesauri, and a few semantic assets focused on Construction circumstances to help in correspondence and comprehension between the various stakeholders involved in a project (Wilson et al., 2001: 85). The Product Libraries space is an unusual example of the confluence of a variety of public and local semantic assets. In any case, these drives will be primarily national in nature, with little regard for the multi-social and multi-lingual nature of the region (Aspin et al., 2011: 152). More recently, the study locale has turned to concerns of collaborative work across a huge number of organizations frequently engaged with a development project with the end objective of working on disciplinary interchanges, smoothing out the store network, and improving consumer relations (Kalay, 1999: 16; Rezgui et al., 2001: 5). The development field has demonstrated some ability to make use of the collaborative developments that have risen to the fore with the rise of the internet. Project extranets, for example, have become a standard tool for larger development projects, while innovations like email are as common in development as they are in most other fields. More widespread acceptance of community-oriented improvements, but still a sluggish interaction when compared to some other fields (Berick, 2004: 235). For example, in Hong Kong's ongoing "One Island East" project, a structure data model was developed collaboratively using CAD innovation from the aviation industry (CATIA, manufactured by Dassault Systems) (Riese, 2006: 11). According to the designers, such initiatives are currently a unique instance rather than the norm, as revealed by the ROAD-CON project's discoveries. In a larger business environment, the willingness to accept new types of coordinated effort is necessitating a shift in perspective, with organizations being forced to relocate their legacy application frameworks with installed information, skill, and business processes to higher-demand applications that shift the focus from intra-endeavor framework joining to between big business process mix (Van Den Heuvel & Mammar, 2003: 105). These larger commercial ventures are advancing to support a changing business climate in which associations enjoy even more transient powerful virtual unions. Furthermore, achieving reconciliation and, as a result, evaluating advantages is heavily reliant on the "combination of divergent frameworks" rather than the traditional highlight point between faces (Sutherland & Van Den Heuvel, 2002: 60). For a long time, the type of fleeting powerful business relationships mentioned above has been the standard in the development field, and they are critical for the foundation of the company exhibitions (Boddy et al., 2007: 678). As a result, we accept that the development field should accept new sorts of coordinated effort and interaction planned innovation to aid them manage their fracture concerns and work on their communication (Shen et al., 2009: 12). As a result, it is argued that the study's strong focus on information integration, while unquestionably important in terms of the fundamental foundation, has influenced the creation and reception of information technology support in the development sector, missing the mark in terms of its latent capacity.

3.3.3 Continuous Acquisition and Life-Cycle Support

The Department of Defense (DoD) established the Computer-Aided Logistics Support (CALS) initiative in the 1980s to standardize how specialized data should be managed within the division (Boddy et al., 2007: 680). Advances in computer technology aided planning, assembly, and design, prompting protective logisticians to shift from a paper-based specialized documentation framework to one in which specialized data is created, managed, and appropriated in computerized structures (Hedberg et al., 2007: 4). It was hoped that the CALS drive would help the Department of Defense save money on specialist documentation by making it more precise, current, and timely (Andrienko et al., 2007: 840). Initially, CALS was primarily responsible for the coordination of help documentation.

As the benefits of the CALS drive became more widely known, DoD security directors sought to incorporate CALS ideas into weapon framework acquisitions (Eide, 2010: 2). The Computer-assisted Acquisition and Logistics Support drive was renamed CALS in 1988 (Boddy et al., 2007: 840). This more accurately reflected its use in dealing with specialized data linked to weapon framework acquisition. Concurrent Engineering (CE) advancements fueled the CALS push to encompass all aspects of weapon framework procurement, including planning, creation, and coordination support processes (Miranda, 2013: 20). Similarly, advancements in media communications, such as undertaking organizing and computerized data trading conventions, have resulted in more specific paperwork being exchanged between companies (Shim et al., 2002: 116). Electronic trade and electronic information sharing, for example, were quickly associated with CALS. Continuous Acquisition and Life-cycle Support is the new name for CALS. The CALS initiative has grown beyond its roots in specialized documentation and coordination to include CE and integrated business processes (Boddy et al., 2007: 840). It has gained acceptance outside of the DoD and security businesses to become a joint DoD-industry oversaw push. The CALS initiative has also been recognized and implemented by international guard divisions in Canada, Europe, Asia, and Australia (Hardman & Marshall, 2000: 206).

The CALS Industry Steering Group has issued a dream explanation, which can be found below. It was extracted from the "CALS Expo 94 International" Conference and Exposition's seventh annual gathering manual (Liao & Vemuri, 2002: 439). The Vision is for all or part of a single venture (e.g., a single hardware manufacturer and its suppliers, or a consortium of public and private gatherings and the academic world) to be able to work from a common computerized data set, incrementally, on the design, development, assembly, distribution, and overhauling of items (Holler et al.,

2014: 5). The immediate benefits would include significant reductions in item to-advertise time and costs, as well as significant improvements in quality and execution (Vereecke & Muylle, 2005: 7).

3.3.4 Project Management Information System

The Project Management Information System (PMIS) is a set of tools and procedures used by project managers to communicate data (PMK, 2021: 1). Undertaking administrators use technological and manual methods and equipment to collect, consolidate, and disseminate data (Keup, 2019: 1). Upper and lower executives use the Project Management Information System (PMIS) to communicate with one another (PMK, 2021: 1). The Project Management Information System (PMIS) aids in the planning, execution, and completion of the executive's goals (Siles, 2014: 6). Project directors utilize PMIS for spending plan structure, such as cost assessment, during the planning system (Anumba et al., 2008: 79). In addition, the Project Management Information System is used to create a specific schedule and define the extension gauge. The project supervisory crew accumulates data into one data set during the execution of the work that the board prioritizes (PMK, 2021: 1). The PMIS is used to compare the benchmark and actual achievement of each movement, to manage materials, to collect financial data, and to save a record for announcement purposes (Lee & Yu, 2012: 84). The Project Management Information System is used to audit the objectives at the end of the project to see if the errands were completed (Siles, 2014: 13). Then it's used to create a final report for the task's completion. The project the board data framework (PMIS) is used to create timetables, budgets, and execute tasks in project management (Kostalova et al., 2015: 98).

The stated information that is created is the most visible piece of data within the management structure. Staff from the project, partner organizations, donors, and recipients construct assessments and make decisions based on the information provided by the task (Ilyas et al., 2013: 3). As a result, a task should aim for ideal, thorough, and reasonable detailed components and organizations (Raymond & Brgeron, 2008: 217). Timing, precision, and fulfillment may be the most important aspects of interior detailing. Everyday tasks rely on a consistent flow of reliable data. A monthly review of project progress and basic issues may be required by the project manager in order to make scheduling decisions (Caniels & Bakens, 2012: 165). Reports are required by nation directors and program supervisors to compare plans and actuals in order to ensure that budgets are not overspent. An organization should devote time and resources to preparing its employees to read internal reports and archives in order to carry out their duties (Ahlemann, 2009: 21). Outside announcing is just as important as inner disclosing in terms of perfection, precision, and fulfillment. In any instance, the configuration or manner in which data is introduced proves to be far more important (Love & Irani, 2003: 651). A giving office may specify the organization, such as the configurations required by the United States Agency for International Development (USAID) (Mignerat & Rivard, 2003: 650). A project manager should make cautious arrangements to report in the required manner in this case. Announcing in the recommended manner may include activities such as programming modifications, staff training, and even innovative changes (Braglia & Frosolini, 2014: 21). The most important consideration for outer details without a directed design is what data the job wishes to pass on and to whom it is providing the data (Siles, 2004: 13). A presentation to country office chiefs about an undertaking's current progress, for example, can be more specialized than a similar presentation to recipients (Yen et al., 2008: 396).

Data management is an important aspect of a task's overall administration. The ability of a project to provide precise, timely statistics to a variety of audiences may have a significant impact on the project's interaction with its employees, board, project recipients, and outside subsidizing organizations (Zika Viktorsson, 2006: 387). Furthermore, the quality and clarity of the data provided to benefactors and other legislative bodies can have an impact on the funding of a work, the legitimacy of a project, and the public's perception of a project (Gelboard et al., 2002: 463) A project data management strategy should be developed to show how data is created, gathered, updated, announced, and stored (Braglia & Frosolini, 2014: 22). This approach, as well as the project's data management system, should be designed to answer two key questions (Keup, 2019: 4):

- Is data being collected, processed, and stored?
- Is data being properly accounted for and conveyed?

Every project has a data management structure; it could be informal and undocumented, or formal and documented. Within the executive's framework, which includes all of an undertaking's data demands, task data management should be unified into the project's extensive data (Lee & Yu, 2012: 85). The data and management structure for a project should handle the use of both mechanical and human resources (Hendrickson, 2003: 392).

3.4 Challenges of Construction Information Management

In the twenty-first century, data management is an essential component of information-based enterprises. Data management enables organizations to be more productive by allowing data to be shared throughout the firm (Penn & Pennix, 2017: 25). This encourages a group to work together, with each member approaching the information base of others (Dibbern et al., 2004: 25). Data administration isn't always without problems. On creating destinations, a few testings, design, and management challenges occur. These issues have an impact on time, finances, planning, and details, and they frequently result in deformities, queries, and postponements. Many building site ideas and practices have remained unchanged throughout time, and identical blunders are being repeated. The main causes are that site management is characterized by high labor over-burden, long working hours, and numerous conflicting gatherings to handle, including the administration of hired sub-workers and consumer contact (Griffith & Watson, 2004: 27). The challenges identified within site management practices, for example, can be divided into three categories: board and organization concerns, specialized and designing issues, and site correspondence issues (Kaliba et al., 2009: 523). Most construction companies have agreements in place that set up methods for the site administrator to be aware of executive and organizational difficulties (Abdul-azis, 2012: 3). These challenges must be addressed in order to ensure that project goals are met. Furthermore, there is a wide range of requirements that may occur nearby, and site chiefs must be prepared to manage each of them in a precise and productive manner (Mohamed, 2006: 233).

3.4.1 Design

The vast number of projects with various development deficiencies, most of which can be attributed to a lack of coordination across configuration features, is the tradition of this perspective on plan in development (Hall, 2006: 1). Interface coordination, craftsmanship control, and support arranging, linkages between parts, risk the board, conflicts between practical projects and the gathering or between the acquisition strategy and the standard presentation, institutional vacuums, and perceptive holes are a few examples (Tombesi, 2009: 118). Examples can be found in Table 3.1. Many of the people involved still believe the Ara Pacis Museum to be a successful building, as seen in Table 3.1 (Tombesi, 2004: 250, 2009: 119, Tombesi et al., 2008: 5; Kleiner & Buxton, 2008: 58). This is because there is a perception that the highlighted defective solutions are not part of the project's life cycle, but rather part of the industry's unavoidable contingent risk. The building facades in Federation Square are an indirect by-product of a combination that never considered building users' needs, partly due to the sequence of design processes prohibiting it and partly due to the lack of a specific brief (Daly, 2021: 5). The Turin Polytechnic campus and the Turin Athletes' Village could be considered victims of the procurement procedures used for their development, as well as the fragmentation of design oversight responsibilities that resulted (Bottero et al., 2012: 205). Due to an awkward interaction between different sections of procurement design, even the Scottish Parliament building in Holyrood, Edinburgh, has been discussed (Newby, 2009: 308). It does, however, testify to the difference between building design and assembly design, as evidenced by the instances of Simmons Hall and the Kursaal Centre (Arau, 2002: 1; Hughes, 2008: 208). Both the bridge in Venice and the Disney Concert Hall demonstrate that both designers and project managers frequently work with enormous uncertainties, which are sometimes exacerbated by the unpredictability of socio-technical conditions (Toyota et al., 2004: 2944). The management of plan in development goes beyond the itemized categorization of assets, the assessment of options, the monitoring of options, or the appropriateness of interchanges within the venture group, as this set of models demonstrates (PointFuse, 2021: 1). Dealing with the project from a planning standpoint also entails ensuring consistency among the multiple floods of intellectual movement that occur across the development's supply chains, which, in turn, have an impact on the character of the building supplied (D4H, 2015: 1). Construction can be viewed as an interdependent and uncertain process in which crucial aspects can extend outside the project's immediate contractual framework, and hence the network of transactions enabled inside it (Hogan, 2009: 1). The split of plan obligations has socio-social/specialized overtones that are regulated and begin (or end) before project-related liabilities. Explicit agreements, on the other hand, can have an impact on the character of the work done under them by dispersing promising situations for participating that are either favorable for successful coordinated efforts or aware of anticipated obstacles/obstacles (Page, 2021: 1). If one accepts this proposition's disciplinary openness, the lines between design management and project design become blurred (Roe, 2019: 1). Because the acquisition of any structure project necessitates the simultaneous dynamic of several independent plan features within the development field, the design management function necessitates reciprocal links between diverse design processes (Dube et al., 2015: 2). Determining such links based on an understanding and assessment of the project's goals, available resources, socio-technical context, timeline, and rate of change internal to each design dimension is an act of design, or project design (Tombesi, 2009: 132).

3.4.2 Data

The patterns of information principles describe how common information items and data segments should be acquired (Hogan, 2009: 2). Information definitions, normalieed queries, and recognized reaction choices are regular set up concepts that guide trustworthy assortment rehearses (Reddy, 2012: 3). For obtaining and conveying development information, there are currently a number of public and statewide information guidelines in use. These principles are not always widely adopted, and they may be in conflict with one another, which can influence the similarity of information collections (Babaei et al., 2013: 375).

Depending on what is typically important for development execution, different types of administrations may use different guidelines. As a result, the degree and detail of data collection may vary across administrations, making comparisons between administrations or population-level informational indexes difficult (Zhang et al., 2013: 184). Because there hasn't been a coordinated effort between customers (government-wide), contractors, specialists, and other organizations to standardize data collection practices, there is a lot of variation in how data is obtained and documented (Faris, 2019: 1).

Client data collection might take place in a variety of situations and settings where it's difficult to gain complete and exact information, and the amount of data gathered will vary depending on the situation (Ceruti, 2004: 2). In most cases, the person in charge of obtaining information plays an important role in the arrangement of a help. Despite the fact that they collect data as part of their duties, data collection isn't the most important skill they have (Jones, 2018: 3). Emergency or crisis situations, when workers are concentrating on an individual's mystery, or situations

Table. 3.1 Design: projects featuring construction shortcomings of different types

Example	Problems Encountered and Results
Ara Pacis Museum, Rome, Italy, 2006	There were issues here with integrating construction components with varying levels of technological expertise, ostensibly without proper interface coordination, workmanship control, or maintenance planning. The result is a large number of as-built absconds that will be difficult to change, as well as a series of unplanned post-inhabitance arrangements that may interfere with its long-term exhibition (Kleiner & Buxton, 2008)
Australian Federation Square, Melbourne, Australia, 2002	The relationship between the outside limit plan and the office plan was problematic in this case (Wong & Teoh, 2008). In this case, the specialized creation of the envelope and its clash with inside space isn't just a result of the modelers' choices, but also a reflection of how their administrations were secured and partitioned, how occupancy was chosen, and how the brief was defined. Workspaces behind the highly intricate triangle-based outer façade of the main building do not have enough protection from the sound emanating from the square, variations in acoustics, and variations in acoustics (Daly, 2021)
Polytechnic of Turin, Turin, Italy, 2006	Here the utilitarian program conflicts with the turn of events and get together of its produced part frameworks, especially in a circumstance where development is being acquired while spending plan related space arranging is yet to be finished. At the Polytechnic, changes in the building's section of parts of the complex have combined with completion pressures and excessive underbidding under a particular form of design-build integrated contract to generate under par subcontracted built work (Bottero et al., 2012: 204)
Winter Olympics Athletes' Village, Turin, Italy, 2006	Major inconsistencies in the reinforced concrete structure of some of its buildings, owing to the fact that different design stages were assigned to different agencies under different contracts, and that these agencies found it difficult (or not convenient) to coordinate decisions at a time when handover dates allowed technical details to be overlooked (Bondonio & Guala, 2011)
MIT's Simmons Students' Hall, in Cambridge, MA, USA, 2002	The architect and structural engineer's decision to design the building's envelope as a giant Vierendeel truss made of precast waffle-like panels (so that the skeleton would conform to the architectural idea) turned the university dorm into a near-impossible construction project, and the most expensive student structure in America per bed. Finally, the general contractor came up with a solution for the panels by building extensive prototypes that were utilized to create and reproduce the required trade sequences (Gardiner, 2008)

(continued)

Table. 3.1 (continued)

Example	Problems Encountered and Results
New Bridge, Venice, Italy, 2008	Even completed artefacts can reveal coordination flaws (or institutional voids) during their creation. Although it was hailed as a structural marvel, it is now awaiting the addition of a steel-frame, egg-shaped cabin for mobility-impaired passengers because it did not (or was not required to) comply with the inclusive design legislation that applies to public works in Italy at the time of the planning permit (Cavalli et al., 2008)
Disney Concert Hall, Los Angeles, USA, 2003	The external volumes of another example of an institutional vacuum were defined according to the city's requirements for vertical surfaces. The decision to replace the original stone cladding with steel boards for esteem design purposes generated light refraction and heat gain difficulties for the surrounding structures, resulting in the rooftop being partially covered with dark canvas (Toyota et al., 2004)

where a customer's satisfaction may be jeopardized due to a widespread mistake are all examples of situations where some information collection may be restricted (TrueLook, 2019: 3).

Further, at times, associations may not be resourced to offer types of assistance to explicit partners, which can mean there is minimal motivator to further develop the information assortment and transmission when a venture is executed (Joyce & Chaboyer, 2015: 3). These people are in an authoritative setting. Alternately, further developed information assortment on need tasks can assist with building a proof base to accomplish viable and effective correspondence crash (Boton et al., 2015: 49).

Occasionally, adequate facts regarding a construction activity to be completed cannot be decided only on a single piece of information, such as when building in a riverine climate (Kent & Becerik-Gerber, 2010: 2). Wherever this is attempted, the effort is typically met with higher risks of relinquishment and barriers to accessing administrations (Pavan et al., 2014: 643). There may be limited freedoms to validate data with an individual interaction with a helper, meaning that the information obtained at the outset cannot be corroborated (Hergunsel, 2011: 32). Furthermore, the complexity of record-keeping frameworks can change, and the quality of information is frequently contingent on the accuracy of the individual entering the data (Aziz et al., 2012: 5).

3.4.3 *Cost*

Most projects require budget management and control, although project extension and cost overruns are common all around the world (Hall, 2006: 24). This can be attributed to a lack of effective methods for evaluating, managing, and controlling customer needs, project scope, and venture cost (Babaei et al., 2013: 376). Various cost-cutting triumphs on major projects around the world in the request for many millions and billions of dollars have sparked growing concern and contemplation at every cultural level (Oke et al., 87). The Global Financial Crisis of 2008 worsened the problem, and it continues to have a significant impact on project finance around the world as lenders tighten lending rules and refrain from lending to projects with insufficient risk management (Smith, 2014: 487). States, major private sector engineers, agents, and international organizations such as the United Nations, the World Bank, and the World Trade Organization have all become increasingly aware of the need for more effective cost management on construction projects (Jones, 2018: 6). This provides a useful global setting for project cost management, discipline to bring issues to light among these significant establishments on the importance and benefit of bringing in a cost expert or quantity surveyor rather than playing the part played by other experts as a subset of their general jobs (Dahmas et al., 2019: 3147).

Significant roadblocks to achieving this include the lack of a single internationally recognized descriptor for the expert discipline (such as planner and engineer) and the absence of common expert guidelines and skills (such as an ISO standard) (Tombesi & Whyte, 2015: 47). Cost engineers, amount assessors, development market analysts, and project administrators are super specialist professions that provide professional cost advices to construction administrations all over the world (Cerutin, 2004: 6). Quantity surveying is a British profession that is mostly recognized in Commonwealth countries. The term "cost engineering" is commonly used in North and South America, China, and a few parts of Europe. Construction Economist is a term that is used as a substitute for the term "assistant" in several European countries and other parts of the world (Smith, 2014: 485). These three expert titles are not viewed with cost the board benefits to a large part accomplished by Project Managers as a characteristic of their set-up of administrations in various districts, especially in Europe.

Project Controls, a relatively new professional field, has also emerged as a useful descriptor of the role of the expenditure administrator (Hogan, 2009: 1). The principal cost of these callings is similar to the board standards and practices. In any event, the cost of the board isn't a prohibitive factor for these professionals, and few countries require official registration to practice as a specialist in the subject (PointFuse, 2021: 1). As a result, cost the board is frequently made up of a variety of experts who may or may not have the expert specialized qualities and aptitude required for viable expenditure management.

3.4.4 Operation

To grasp activities and how they contribute to an organization's success, comprehend the fundamental concept of tasks, the value added nature of tasks, the impact innovation can have on execution, and the world's competitive commercial center (Ofori, 2000: 17). For clients/customers, productive association activities are a critical tool in gaining the upper hand in the day-to-day problem (Mohamed & Anumba, 2006:

233). What factors influence these elements' purchase decisions? Value, quality, item execution and elements, item assortment, and item accessibility are all important factors for most services and products (Griffith & Watson, 2004: 30). The activities performed in tasks have a large impact on this plethora of variables. For example, as item usefulness increases, item costs decrease, and item cost can be reduced. Furthermore, if better manufacturing techniques are developed, the quality and variety of products available may increase (D4H, 2015: 1).

Cooperative energy can be generated by linking activities and working procedures with the organization's overall strategy (counting, monetary, advertising, and data framework methods) (Kaliba et al., 2009: 523). When offices, equipment, and worker preparation are perceived as a means to achieve authoritative goals rather than as barely engaged departmental destinations, activities become a positive component (Mohamed & Anumba, 2006: 235). As a result of this shifting perspective, the rules for deciding on tasks are shifting away from cost management (a vaguely defined working aim) and toward global execution estimates in areas such item execution and assortment, item quality, delivery time, client service, and functional adaptability (Smith, 2014: 492).

Mechanical information is an important aspect of functional flexibility in many businesses in today's business climate (Page, 2021: 1). With advancements in technology, it is now possible to produce better products with fewer resources. Because innovation fundamentally alters an item, its presentation and quality typically improve, making it a far more highly valued product in the marketplace (Chen, 2019: 4). However, advancements in highly advanced business apps have spawned new competitors, necessitating organizations' efforts to enlist benefits in all feasible areas of task management (Chron, 2021: 1).

After some time, the board's operations have broadened in scope and significance (Stewart, 2006: 3). It now has essential components, relies on conduct and design ideas, and employs executives' science/activities research gadgets and methods for exact direction and critical thinking (Altameen et al., 2014: 22). As the executives continue to develop duties, they will increasingly link with other practical areas inside the organization to encourage integrated approaches to complicated interdisciplinary concerns (Hoezem et al., 2006: 4). To be sure, such collaboration is often seen as critical for private firm foundations and global organizations alike to achieve long-term corporate success (Ofori, 2000: 25).

3.4.5 *Personnel*

Throughout the course of a typical day, people are confronted with the need for large and small amounts of data (e.g., "When is my next gathering?"; "What's the status of the financial plan speculation?" "What's in the news today?"); stimulating workouts to find and find again (Whittaker, 2011: 6). A large number of investigations in data chasing, data conduct, and data recovery are focused on attempting to locate data in public places such as the internet or a traditional library (Teevan et al., 2004: 450).

Even in efforts to track down fresh info, previously unexperienced, from a public source like the web, there is a strong individual component (Whittaker et al., 2011: 151). For example, data collection efforts could be organized by a custom-made plan, a self-addressed email update, or a daily agenda. Furthermore, information contained in a person's PSI can be used to aid in a more targeted, tailored web search (Whittaker, 2011: 3450).

Rather than a single transaction, an individual's efforts to observe meaningful facts are usually a series of communications. Data is acquired in chunks through a series of communications in a "berry picking" model of discovery, and throughout this time, an individual's demeanor of need, as indicated in the current inquiry, grows (Jones et al., 2010: 2804). Individuals may prefer to deal with tracking down essential data in a sequential manner in order to maintain a greater sense of control and setting over the process. More modest advancements may also reduce the cognitive burden associated with inquiry outlining (Teevan et al., 2004: 416). There are times when there isn't a quick way to get to the information. For instance, an individual's recognition of a required site could be based solely on an email message sent by a colleague, i.e., an individual may not recall a web address or even catchphrases that could be used in a web search, but the individual recalls that the site was mentioned recently in an email from a colleague (Jones, 2007: 455).

Individuals may find (rather than re-find) data in any case, when this data appears to be heavily influenced by. Things could be pushed into the PSI (for example, via the inbox, online recording memberships, and downloads) (Fitchett et al., 2013: 2330). If these items are later discovered, it will be through a demonstration of finding rather than re-finding (since the individual has no recognition for the data). Lansdale described data recovery as a two-step process including interaction between review and perception processes (Czerwinski et al., 2004: 176). The methods of evaluation and acknowledgement can be repeated indefinitely to minimize the efforts to locate the ideal data. This transaction happens when people go through an envelope order to find the perfect record or email message, or when they go through a website to find the perfect page (Bergman et al., 2013: 1996).

In any case, re-finding begins with a different sequence; make sure to explore the initial location first (Mann et al., 2004: 4). Individuals may go to great lengths to create online bookmarks or archive material, only to discard it later, resulting in, in the most pessimistic scenario, a waste of the initial effort (Bergman et al., 2013: 3). Similarly, finding or re-finding frequently entails gathering a single piece of data as well as a large amount of data (Bergman et al., 2009: 270). It's possible that the individual will have to repeat the search sequence a few times. A test in device assistance is to provide persons with tools to collect or interrelate data items such that their chances of recovering a complete arrangement of the data required to complete a task improve (Jones et al., 2010: 3590).

In the long run, PIM has not truly established that people prefer to explore rather than hunt for personal data, particularly data stored in near to home computerized records (Dinneen & Frissen, 2020: 2). Support for browsing through data near to home has improved greatly over time, most notably in the arrangement for full-text ordering to improve search speed (Dumais et al., 2003: 73). With these enhancements, inclination may be on its way to being a necessary approach for locating email messages (e.g., search on subject or sender, for messages not in view).

Nonetheless, a preference for route as the primary way for locating close-to-home records (e.g., stepwise envelope crossing; reviewing a list of documents inside an organizer for the right document) persists, despite ongoing improvements in search support (Alon & Nachmias, 2020: 144). The suffering preference for route as a necessary means of returning to records may have a neurological basis, in that route to documents appears to utilise psychological facilities similar to those used to explore in the real world (Dumais et al., 2003: 73).

The preference for route is also in accordance with a dominance effect that is frequently observed in mental examination, to the point where the preferred method of return aligns with the initial openness (Jones et al., 2010: 2810). If an individual's underlying involvement in a document recalled its situation for an envelope, where the actual organizer was reached by exploring through a progressive system of containing organizers, then the individual will incline toward a comparative technique for route for later return to the record (Whittaker, 2011: 57).

There have been some prototyping initiatives to examine an in-setting creation, for example, development of records and other forms of data, such as web references and email, for a venture the individual is working on (Jones, 2007: 3590). Prototyping efforts have also looked into ways to improve route assistance, such as by highlighting and making it more straightforward to follow the ways people are likely to return in their route to a document (Alon & Nachmias, 2020: 144).

3.5 Benefits of Construction Information Management

The management of information assets is gaining traction around the world. Chiefs are (or should be) beginning to realize that by ignoring one of the four assets they have control over (along with monetary, human, and physical resources), they are limiting their own and their employees' capabilities for a wide range of tasks (Price, 2015: 1).

Data is the lifeblood of any company or establishment. It aids in the direction of how organizations structure techniques and implement processes that rely on them (Jones, 2021: 3). It is at the heart of business development, which is why so much effort and money is poured into developing effective data the board frameworks and experienced specialists to aid with their implementation (Kier, 2018: 10). For businesses, increasing productivity and gaining a competitive advantage means increasing profits.

Furthermore, there is definitely another side; with access to vast amounts of data comes enormous responsibility (Edology, 2014: 1). We routinely exchange data with associations, both voluntarily and involuntarily. Associations are under increasing pressure to handle data responsibly and ethically, as well as to follow evolving legal rules (Bryde et al., 2012: 3). Individuals and states are demanding more transparency, therefore dealing with the data about individuals that associations have is a higher priority than it has been in recent memory, and the need for competent, safe, and sustainable Management Information Systems is greater than it has ever been (Harris, 2018: 1). Failure to follow these requirements can find businesses in hot water, with potentially crippling fines, which is one of the reasons why businesses take data management so seriously (Bruton & Lynn, 2015: 2).

3.5.1 Communication

Correspondence and management are inextricably linked. Correspondence refers to the process by which data is sent between at least two people; robots are increasingly being employed for correspondence, but we'll stick to individual correspondence for now (Wickforss & Lofgren, 2007: 338). Every aspect of administration, including planning, sorting, driving, and controlling, is dependent on effective communication (Ackerman, 2000: 181). Supervisors should be able to obtain precise data in order to make plans, and they should also be able to send precise data for the designs to be implemented (Nitithamyong & Skibniewski, 2004: 495). Everyone in an organization can be educated when data is sent and received precisely. However, as we saw in the last model, when data is jumbled or incorrect data spreads, interchanges can cause major problems in organizations (Rezgui & Zarli, 2001: 83).

The job of management is to help an organization achieve its goals. To do this, directors create a plan that specifies what should be done, when it should be done, and how it should be done (Price, 2015: 1). Administrators should share this information with everyone in the organization in order to carry out the plan. That is, they should inform members of the organization about the plan. Chiefs, on the other hand, must do a lot more than only instruct people on how to support the arrangement (Bruton & Bruton, 2015: 2). They should also encourage people to help the plan, create a sense of commitment to the organization, establish compatibility and cooperative effort, and keep everyone informed about events and activities that affect the organization (Chen & Kamara, 2008: 5). Great correspondence illuminates and creates a culture that helps people feel like they belong and want to serve the organization. The following are some of the advantages of effective correspondence (Bruton & Bruton, 2015: 2).

- **Provides clarity**: Confusion, vulnerability, and ambiguity make people feel uncomfortable and uncooperative. Making roles, duties, and links explicit offers everyone the information they need to carry out their responsibilities and understand their responsibilities to the organization. Successful correspondence reduces the costs of miscommunication, errors, and omissions.
- **Builds Relationships**: A culture that encourages open communication reduces tension amongst personnel at various levels, both expertly and socially. Individuals are compelled to seek assistance with concerns and to recommend arrangements and upgrades in a trustworthy and group society. Collegial culture is

fostered by effective correspondence, which encourages collaboration and enables participation.

- **Creates responsibility**: Sending and receiving data are both important aspects of effective correspondence. Chiefs may make everyone in the organization feel like valued donors by paying attention to representatives' interests, allowing them to have input on their work and working environment, and thinking about their ideas. When employees feel valued in the organization, they are more likely to be drawn in and inspired. Backing and responsibility come from effective correspondence.
- **Defines assumptions**: When people are unsure about what is expected of them and how they will be evaluated, they are unable to carry out their obligations effectively. Execution audits are difficult because the representative has no understanding what exhibition principles they are expected to satisfy. Furthermore, if corrective action is required, the representative may become enraged if he is unable to comprehend how his actions harmed his viability. Workers know what they need to do to get a good survey and the benefits that come with it when assumptions and principles are clear.

These are just a few of the numerous benefits that come with good interchanges. When everyone in the organization is focused on the goals, supervisors may be able to achieve hierarchical objectives. When people are educated and involved, they do significantly better (Peansupp & Walker, 2006: 365).

The correspondence interaction may appear simple: one person delivers a certain message, and others receive it (Jewell & Walker, 2005: 113). Despite this, the cycle proves to be more perplexing due to the fact that the data in the message must be transmitted and received precisely (Koskele & Kazi, 2003: 63). The correspondence interaction model depicts how data is transmitted and received. This model is depicted in the accompanying diagram (Fig. 3.2).

When one person is conversing with another, the model is easiest to comprehend. The person who initiates the correspondence, the sender, has information that the recipient, the recipient, needs to know (Koskela & Kazi, 2003: 63). Nonetheless, the data must be encoded into a structure that can be communicated before it can be sent (Love et al., 33). In a simple scenario, the information is well-articulated and directed to the recipient. Alternatively, the data could be converted into printed texts, tables, diagrams, or charts and distributed to the receiver (Peansupp & Walker, 2006: 195). In a more complicated scenario, the data is encoded into words or pictures, which are

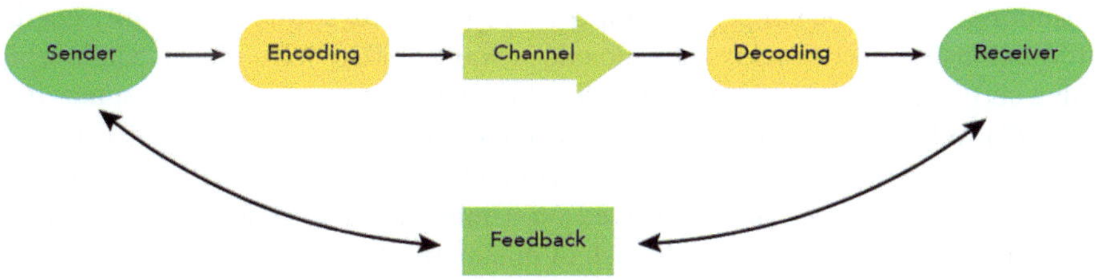

Fig. 3.2 The communication process model. *Source* Bruton and Bruton (2015: 2)

then converted into electronic signs and sent to the intended recipient (Christensen & Walker, 2004: 41). The medium over which data is transmitted is referred to as the channel (Maqsood et al., 2003: 615). It could be air traveling through solid waves, paper transmitting text and images, or cables or attractive fields transmitting electrical signals (Marsh & Flanagan, 2000: 424).

The beneficiary might flip the interaction to further clarify the model. He receives the encoded message and then decodes it. That means he converts the message into data that can be understood again. In the first model, a worker reads the message and determines who has been hired and when he will start work. Supervisors have handed over data to workers (Peansupap & Walker, 2005: 257). In an intelligent correspondence procedure, the recipient might provide feedback to the sender to demonstrate that the message was received and interpreted (Peansupap & Walker, 2005: 137). This can start a natural back-and-forth exchange, assuring the sender that their message has been received and seen (Christensen & Walker, 2004: 41).

3.5.2 Point of Contact

A point of contact (POC) is a person or office that handles all inquiries and solicitations. They will be in charge of daily correspondence, executives, and the swift resolution of administrative demands and claims (Wang & Huang, 2006: 255). In most cases, there are multiple ways to contact a single resource, such as a phone number, email address, website, or application (Westerveld, 2003: 413).

For improved communication, each customer is assigned a single Personal Information Manager (PIM) for all services (Yu et al., 2005: 430). Aside from the strong customer and worker entryways, each company also has the responsibility of having only one point of contact so that the consumer can build a relationship with the assistance representative (Shenhar et al., 2001: 670). Instead of a single point of contact, the customer is aware of the Personal Information Manager's identity, as well as several contact methods such as email, telephone, and customer entryway management tickets (PMBOK, 2004: 10). Customers' experiences with a single resource in terms of connection and administration have a big impact (Chen et al., 2005: 9).

No Need to Re-explain

Finding the right person to handle your problem can be frustrating and time-consuming. When you simply have one help representative, they can usually get the information they need quickly and efficiently (Wang, 2003: 4). Building a relationship with a certain agent supervisor is beneficial to everyone, not just the consumer (Qi, 2002: 72). When there is only one designated individual data administrator, everyone can be more useful. Because there's no compelling reason to clarify anything, all enquiries and requests can be resolved on a mutually agreeable basis (Liu et al., 2003: 205). A SPOC eliminates the risk of misinterpretation and eliminates the need to repeat questions and demands over and over (Yoon et al., 2003: 70). Data has never been so swift or adaptive, because to the organization's incredible customer and representative entrances. All correspondence is coordinated by SPOC, ensuring that nothing is lost or misplaced at any time (Warkentin et al., 2000: 85).

No Wasted Time

Customers understand the frustration of calling a company's employees and getting different answers from different people (Cao, 2002: 10). Significant wait times, inexperienced employees, and phone anchors appear to be the norm. These events are not only unsatisfactory, but they also waste a lot of time. It's rare that any development problem can be solved in a single day. When precision and practicality are required, a SPOC is essential (Cao et al., 2006: 299). Supervisors can also have immediate access to all of the organization's and representatives' assets, as well as a dedicated individual data director to deal with anything that arises (Roth et al., 58).

Expanded Trust and Dependability

When the PM or customer can rely on a genuine individual for all of their enquiries and requests, a better relationship will be developed among all parties involved (Cao et al., 2007: 7). Your involvement in a PEO becomes more positive and enjoyable when you have a single resource. The individual data supervisor is in charge of the day-to-day correspondence as well as the HR, risk and consistency, benefits, and financial demands of your company (Chiou & Lee, 2002: 1499).

Viable and Efficient

When you can phone or email your records administrator directly, you'll never have to wait on hold or be transferred from one department to another (Chiou & Lee, 2005: 812). Rather of wasting time worrying about your HR, benefits, finance, safety, and consistency requirements, you may focus on what matters most to you. The following are some of the advantages of staying in touch (Sydorenko, 2021: 1):

- The resource is the face of the organization and is in charge of grievances and requests from clients and customers. The organization has a website, and you will frequently see people looking for information or complaining about certain products or services. The resource easily answers each query, ensuring that the firm does not have to deal with the wrath of any disgruntled customers.
- When a customer walks through the door of a store, the person who answers his questions acts as a resource for him. The resource between you and the café is the representative who will come to make you agreeable and encourage you to hang tight for a couple of moments when you enter an eatery and don't find a vacant table.
- He occasionally goes above and above to assist customers and clients in general by providing premium administrations. For example, while I was waiting for a table at a rumored restaurant, my niece started weeping out of nowhere. The resource appeared and gave her a treat to eat, which she accepted.
- Because he is an organization's known substance, the resource remains in the spotlight. He should establish a positive commitment to the client and assist him.

The rapid spread of news has been facilitated by online media and interconnection. They associate his proper behaviour with a brand image when the resource conducts cheerfully and enjoyably with the clients.

- Assume you're in an airport and you can't see your ticket. The person who guides and assists you becomes your resource at that point. Through web-based media stages and mouth exposure, clients exchange pleasant and negative feelings; as a result, a resource should be cautious in his responsibilities.
- He is eager to build a practical relationship with clients and responds to each question enthusiastically. He is prepared to resolve any issues that may arise in order for the client to have a positive experience. Customers regard it as trustworthy because they can easily communicate their needs and concerns to him. A resource's efforts have a long-lasting impact on the personality of his clients.
- For example, a client was having some issues with his bank's archives. When the resource noticed this, he arrived right away to provide his services, making the client's experience stress-free and enjoyable. The client truly appreciated and acknowledged his administration.

3.5.3 Operational Efficiency

Functional productivity means that labor and products are produced or delivered faster, cheaper, better, fresher, with fewer flaws, and so on, thanks to the IoT's caution (Uusitalo et al., 2019: 3). Although the functional effectiveness improvements for a certain IoT application or framework may be so great that this is the primary requirement and protection for a venture, client satisfaction is always important (Bilal et al., 2019: 4).

Functional efficiencies will account for a large portion of the advantages in IoT and development situations, and thus efficiencies indicate that administrations are delivered with less expenditure, fewer people, and fewer assets consumed (Alizadeh & Yitmen, 2018: 5). In most cases, operational productivity will be the driving force behind the IoT framework or administration in question, particularly if the support is affiliated with non-profit entities such as government or related medical services (Lee et al., 2018: 123). In any event, functional productivity is a positive thing in development companies, and it will undoubtedly promote IoT frameworks for reasons such as having a competitive advantage (Marchi et al., 2018: 3).

Functional effectiveness is measured in terms of adaptability, consistency, utility, and process time, and is centered on improving internal worker exhibition (Chang et al., 2018: 5). As business conditions change, a business cycle's adaptability is defined as the ability to address immediate and long-term client needs (Rakhmatullina, 2017: 2851). Data improvements, such as PC-assisted assembling and gathering, enable businesses to adopt a more adaptable way of assembling and increase client responsiveness (Kudryavtseva et al., 2015: 267). IS also aids firms in reacting to changing economic conditions by providing faster and more exact data about customer wants, costs, and expenses (Gareev & Mukhametova, 2018: 307). Adaptability is defined as the ability to fundamentally increase or decrease a restriction without causing major disruption or incurring excessive costs (Walasek & Barsszcz, 2017: 1230). In view of the fact that non-adaptable cycles demand enormous capital consumption for creation framework limit alterations, one of the substantial advantages of specialized improvements in data advances is the expansion of versatility (Matejka & Tomek, 2017: 1081).

In a business cycle, consistency refers to how comparable desired outcomes are obtained by using similar techniques (Salem et al., 2018: 3431). By combining virtually all ways in frameworks, IS may drive organizations to perform things reliably. IS could also be used to provide data that helps people perform and control business activities in a predictable manner (Kokliugina & Kokliugin, 2016: 229). When it comes to business processes, efficiency is defined as the relationship between the yield generated by a business cycle and the amount of money, time, and effort it consumes (Borovskikh, 2017: 3395).

The ability to generate extra yield from equivalent degrees of data sources or provide similar yield from lower levels of data sources is critical for improving the usability of business processes (Anysz & Rogala, 2019: 184). IS helps employees work smarter and more efficiently by providing the proper data to allow workers to skip steps in task handling, robotizing tasks and calculations, assisting with autonomous guidance, and assisting with creation handling and planning (Borovskikh, 2019: 15). It's important to remember that IT projects include a wide range of costs, including assistance, planning, support, organization, and time spent by end-users programming (Shagiakhmetova & Serdarova, 2018: 86).

The time between the start of an engagement and its completion is referred to as process duration. It is influenced by a combination of three factors: handling time, holding time, and conditions between stages (Gilyazutdinova et al., 2016: 75). It was calculated that in confused assembly with plants producing a wide range of items, less than 10% of the total process time had been consumed, while the material was being improved at the same time. IS can aid in situations where there are large delays and bottlenecks by assisting individuals in deciding on the best task request to improve the work process. Different drawing equipment, factual examination gadgets, and nitty gritty data are used to refine this (Zawistowski & Kulejewski, 2018: 563).

3.5.4 Cost and Schedule

During the building cycle, the most important goal is to complete the project on time and within budget while adhering to established quality standards and other details (Al-Hazim et al., 2017: 19). To achieve so, a strong focus on dealing with the construction cycle is required. In any case, dealing with a construction cycle is impossible without a plan and a system in place (Ameh et al., 2010: 51). The goals for a project's timing, cost, and asset utilization, as well as the assignments and techniques for executing the task, are all spelled forth in an agreement (Safapour et al., 2017: 3). The organization is usually based on previously recorded data sets and previous participation in comparison research (Mohamad et al., 2012: 2).

A control framework, on the other hand, gathers genuine information (input) on a project's timetable, cost, and asset use; considers current progress to the planned timetable (investigation) to identify potential pain points that require special attention; and makes decisions based on the results of the investigation (Liao et al., 2011: 230). A coordinated cost control and timetable control task is expected to support the development process the board viably gather quality information in an ideal manner and provide quality recorded information bases for future preparation of new ventures (Kermanshachi et al., 2017: 149). A substantial percentage of development projects employ some type of cost and schedule control method. Overall, many activities suffer from the negative effects of ineffective control due to the wasteful progression of data (Marzouk et al., 2008: 317). This shortcoming is a fundamental problem in the growth of executives. The quintessence of the problem is the type of data streaming in the control structure (Le-Hoai et al., 2008: 368). Issues are discovered both in the early stages of a project's development and in the corporate archived information base. Incorporating cost-and timetable control works, as well as regulating information quality, trustworthiness, and idealness, are two challenges that must be overcome in order to find a solution to this problem (Said & El-Rayes, 2010: 423).

Timetables and costs are intricately linked in construction projects. Although the interdependence of time and cost is self-evident, project-management frameworks that include expense and timetable control capabilities are unusual (Yeo & Ning, 2006: 124). Rather, they remain two distinct capacities that operate independently of one another and employ two distinct control structures: the work breakdown structure (WBS) and the cost breakdown structure (CBS) (Shane et al., 2009: 222). The difficulty in integrating the two capacities stems from the level of detail that each uses rather than the level of detail that is necessary. The CBS's expenditure control work is done at a lower point-by-point breakdown level than the WBS's timeline control effort. Because of the difference in the level of detail used by each capability, there is a significant difference in how cost and schedule information is kept up to date. Every collection of data becomes independent of the others and is kept up with on its own. The data from two sources need then be correlated, which reduces the effectiveness of obtaining meaningful data.

To achieve the appropriate mix, cost and time data should be obtained and maintained at a set up shared factor control account described at a sufficiently detailed level (Dao et al., 2017: 8). Nonetheless, establishing and maintaining a common component is insufficient. For safeguarding and storing information that supports this principle and amount of detail, a robotized technique is also required (Kamalirad et al., 2017). As a result, a calculating information gathering and capacity model that preserves the coordinated expenditure and schedule control concept and provides quality information in a straightforward manner is required (Kermanshachi, 2016: 35).

3.5.5 Market Insight

There is nothing quite like fantastic and comprehensive data, and nothing quite like exhaustive information (Bhatia & Awasthi, 2018: 1036). The most basic reason is that nothing is static, because every human on the world makes daily decisions and performs innumerable demonstrations based on fragmented facts combined with a specific level of limited information (Jung et al., 2008: 623). We recall the goal of marketing as a science, the establishment of common norms, and the revelation of methods for achieving awareness of opportunities (Bolivar-Ramos et al., 2012: 333). Market knowledge, it is stated, is one of the overall structure blocks essential for a sending out firm to deliver need-fulfilling things to customers (Fang et al., 2016: 69). Market Insight is also a concept that combines the element of being aware of possibilities. As a result, Market Insight is a two-dimensional concept, one of which may be precisely upheld, while the other, which is associated with recognizable proof of chances, can only be properly upheld (Ganesan et al., 2010: 3630).

The term understanding has several meanings that are nearly indistinguishable. In this particular circumstance, the following one is chosen: the capacity to see clearly or deeply (Herdian, 2014: 5); infiltration, an entering and frequently abrupt arrangement that begins with a perplexing situation (Ilkay & Aslan, 2012: 754); or issue Psychology (Setiawan & Hanfan, 2017: 247), the limit in terms of understanding one's own psychological cycles and the prompt comprehension of the meaning of an event or activity. In this regard, the ability to observe and quickly comprehend the meaning of an event are deciphered to capture the element of having the choice to deliver need-fulfilling things to customers, while brain science and the ability to comprehend one's own capture the aspect of being aware of opportunities (McGuinness & Morgan, 2005: 1308). This is an important distinction in this proposal since it defines the ability to deliver need-fulfilling things to customers as a very straightforward interaction (Prajogo et al., 2012: 308). One has an item, it is sent out, people receive it, and then, based on data, one alters and adapts marketing strategies and the item based on standard input. This is the same old thing, and it could be proposed exactly (Sun & Ni, 2012: 873).

The element of being aware of possibilities is distinct in that it describes a void in this specific situation; it may be a market segment not yet identified by others, a need not addressed by any other item accessible, and so on (Phan et al., 2011: 520). The advantage of being able to choose between the two is clearly superior (Mehra & Coleman, 2016: 1126). Market insight is a fact that hasn't been utilized in a long time, but when it is, it will generate profit. These realities or opportunities exist simply because no one has access to all data and information is restricted among all competitors (Switzer & Bourdon, 2011: 135). The Black Swan hypothesis is an excellent example of what this entails. In the context of this review, this definition has been deciphered to the point where there are a plethora of facts, Black Swans, and unidentified freedoms on the lookout, and they simply exist, waiting to be discovered and utilized by the organization that recognizes their genuine presence and organizes itself accordingly; that is, to see them and get a handle on them (Trentin et al., 2012: 852). In this context, handle denotes both psychologically and functionally.

Experience and information are the primary building blocks for developing Market Insight, with the size of relevant organization relationships acting as the arbitrator (Kafetzopoulos et al., 2015: 383). One could hypothesize that the likelihood of getting market insight(s) increases proportionally with the number of company contacts who are willing to share their knowledge and experience with the company's executives. One may argue that these are basics of market direction and internationalization speculations, as well as data behavior (Yunis et al., 2013: 692). The subsequent aspect, however, the nonlinear or awry revelation aspect, which this theory combats, isn't caught by either of them; thus, the bi-dimensional concept of Market Insight addresses a crucial aspect for satisfying the points of the study of Marketing in a sending out setting in the twenty-first century (Setiawan & Hanfan, 2017: 247). The challenge, however, is that these chances can't be found by using insights or applying deductively established basic conduct norms. As a result, no attempt has been made to quantify this part of the investigation. In any event, such possibilities or certainties cannot be identified using the logical method (Kaynak, 2003: 407).

3.6 Lessons Learnt

The literature survey for this part revealed that projects in the development industry are becoming increasingly large and complex, with new development breakthroughs, methodologies, and so on rapidly expanding (Lee et al., 2018: 1). The study revealed that by combining these members in various ways, few variations of the authoritative hierarchy for CEOs are achievable in an optimal situation. For example, there is a cycle in which the overall worker for hire signs a total agreement with the customer, followed by the worker for hire signing contracts with the subcontractors. In another cycle, the client may sign an agreement with the worker for hire for the main components of the task, and then sign the excess pieces of the venture with the sub-workers for hire. Furthermore, developers are establishing increasingly divergent expectations as all related laws and norms change rapidly. Because of these advancements in development innovation and the growing complexity of the development industry, effective management of the many data generated by construction projects is becoming increasingly important. Specifically, because the development stage is where the structure begins to take shape based on the blueprints, an effective task necessitates comprehensive data management in order for the development stage to work. Because data is so important to executives, the construction industry employs a variety of IT strategies, investigations, and techniques, such as Computer Integrated Construction, Continuous Acquisition and Life-cycle Support, Project Management Information System, Traditional data management, and Building Information Modelling.

In any event, the literature review revealed that data management is rarely problem-free. On constructing sites, there are a few testing, designing, and board concerns. These difficulties have an impact on time, budget, planning, and details, and they frequently result in flaws, disagreements, and delays. Many building site ideas and practices have remained unchanged throughout time, and same blunders are repeated. The main reasons are that site management is characterized by excessive job over-burden, extended working hours, and numerous conflicting gatherings, such as sub-project worker administration and client's contact. For instance, concerns identified within site the board practices can be classified into three categories: executives and organization issues, specialized and design issues, and site correspondence issues. Most construction companies have procedures in place that outline how the site supervisor can keep track of board and organization issues. These difficulties must be addressed in order to ensure that the project's objectives are met. Furthermore, there is a wide range of imperatives that may occur in the immediate vicinity, and site administrators must be prepared to manage each one wisely and successfully.

The advantages of construction data for organisational hierarchies were also examined in this area. Data is the lifeblood of any company or organization. It guides how businesses establish procedures and carry out operations that rely on them. It is at the heart of business development, which is why so much effort and money is put into developing effective data the board frameworks and experienced specialists to help with implementation. For businesses, increasing productivity and gaining a competitive advantage means increasing profits. There is, of course, another side to this; with access to vast amounts of data comes enormous responsibility.

References

Abdul-Azis, A. (2012). Construction project performance evaluation using integrated project management methods. *International Journal of Project Management*, *30*(4), 456–468. https://doi.org/10.1016/j.ijproman.2011.10.006

Ackerman, M. S. (2000). The intellectual of CSCW: The gap between social requirements and technical feasibility. *Human Computer Interaction, 15*(2–3), 179–203.

Adekunle, P., Aigabvboa, C., Thwala, W., Akinradewo, O., & Oke, A. (2022). Challenges confronting construction information management. *Frontiers in Built Environment, 8*, 1075674. https://doi.org/10.3389/fbuil.2022.1075674

Ahlemann, F. (2009). Towards a conceptual reference model for project management information systems. *International Journal of Project Management, 27*, 19–30.

Al-Hazim, M., Al-Ahmari, A., & Al-Amri, S. (2017). Risk assessment for construction projects using advanced probabilistic models. *Journal of Construction Engineering and Management, 143*(8), 04017052. https://doi.org/10.1061/(ASCE)CO.1943-7862.0001319

Alon, L., & Nachmias, R. (2020). Anxious and frustrated but still competent: Affective aspects of interactions with personal information management. *International Journal of Human-Computer Studies, 3*(1), 144.

Alizadeh, A., & Yitmen, I. (2018). Information and knowledge management in construction projects: A review. Automation in Construction, *93*, 1–12. https://doi.org/10.1016/j.autcon.2018.05.001

Altameen, A., Alshibani, A., & Mohamad, R. (2014). Project management information systems and construction efficiency. *Procedia Engineering*, *77*, 115–123. https://doi.org/10.1016/j.proeng.2014.07.023

Ameh, O., Ahiaga-Dagbui, D., & Osei-Kyei, R. (2010). Risk management in construction projects: Towards an integrated approach. *International Journal of Project Organisation and Management*, 2(2), 124–144. https://doi.org/10.1504/IJPOM.2010.032096

Andrienko, G., Andrienko, N., Jankowski, P., Keims, D., Kraak, M. J., Macheachren, A., & Wrobel, S. (2007). Geovisual analytics for spatial decision support: Setting the research agenda. *International Journal of Geographical Information Science, 21*(8), 839–857.

Anumba, C., Pan, J., Issa, R. R. A., & Mutis, I. (2008). Collaborative project information management in semantic web environment. *Engineering, Construction and Architectural Management, 15*(1), 78–94.

Anysz, H., & Rogala, P. (2019). Risk and knowledge management integration in construction projects. *Engineering Management Journal, 31*(1), 45–56. https://doi.org/10.1080/10429247.2018.1559791

Arau, E. A. H. (2002). The acoustic design of the Kursaal Center. *Pacs reference SS-RBA-02: Estudi Acustic H.Arau* (pp. 1–6).

Aspin, R., Dadalte, L., Fernado, T., Gobbrtti, E., Marache, M., & Shellbourn, M. (2011). A conceptual Framework for multi-modal interactive virtual workspace. In *ITcon, 6, Special Issue Information and Communication Technology Advances in the European Construction Industry* (pp. 149–162). http://www.itcon.org/2001/11

Aziz, A. A. A., Memon, A. H., Abdul-Rahman, I., Nagapan, S., & Latif, Q. B. A. I. (2012). *Challenges faced by construction industry in accomplishing sustainable goal* (p. 628). Universitas Bakrie, Jakarta, Indonesia, Faculty of Civil and Environmental Engineering, Universiti Tun Hussein Malaysia.

Babaei, M., & Beikzad, J. (2013). Management information system, challenges and solutions. *European Online Journal of Natural and Social Sciences, 2*(3), 374–381. ISSN 1805-3602.

Bercerik, B. (2004). A review on web-based project management tools and their adoption by US AEC industry. *International Journal IT, Architecture Engineering Construct, 2*(3), 233–248.

Bergman, O., Gradovitch, N., Barllan, J., & Beyth-Maron, R. (2013). Folder versus tag preference in personal information management. *Journal of American Society for Information Science and Technology, 64*(10), 1995–2012.

Bergman, K., Borkar, S., Campbell, D., Carlson, W., William, C., et al. (2009). *ExaScale computing study: Technology challenges in achieving exascale systems*. Defense Advanced Research Projects Agency Information Processing Techniques Office (DARPA IPTO), Tech. Rep 15.

Bhatia, M. S., & Awasthi, A. (2018). Assessing relationship between quality management systems and business performance and its mediators: SEM approach. *International Journal of Quality and Reliability Management, 35*(5), 1034–1059.

Boddy, D., Watson, T., & Paton, R. (2007). The role of project management information systems in construction. *International Journal of Project Management, 25*(7), 707–716. https://doi.org/10.1016/j.ijproman.2007.01.002

Bondonio, P., & Guala, C. (2011). Gran Torino? The 2006 Olympic Winter Games and tourism revival of an ancient city. *Journal of Sport & Tourism, 16*(4), 303–322. https://doi.org/10.1080/14775085.2011.635015

Borovskikh, P. (2017). Information management for construction project risk assessment. *Procedia Engineering, 196*, 457–465. https://doi.org/10.1016/j.proeng.2017.08.050

Borovskikh, P. (2019). BIM and digital collaboration in construction projects. *Automation in Construction, 107*, 102924. https://doi.org/10.1016/j.autcon.2019.102924

Boton, C., Halin, G., Kubicki, S., & Foegues, D. (2015). *Challeneges of big data in the age of building information modelling: A high-level conceptual pipeline* (pp. 48–56). Springer.

Bottero, M., Sacerdott, S. L., & Mauro, S. (2012). Turn 2006 Olympic Winter Games: Impacts and leagues from a tourism prospective. *Journal of Tourism and Cultural Change, 10*(2), 202–217. https://doi.org/10.1080/14766825.2012.683954

Braglia, M., & Frosolini, M. (2014). An integrated approach to implement project management information systems within the extended enterprise. *International Journal of Project Management, 32*, 18–29.

Brown, R. J., & Renshaw, P. D. (2000). Collective argumentation: A sociocultural approach to reframing classroom teaching and learning. In *Social interaction in learning and instruction:The meaning of discourse for the construction of knowledge* (pp. 52–66). Pergamon/Elsevier Science International.

Bruton, G., & Bruton, A. (2015). Knowledge management practices in large-scale construction projects. *International Journal of Project Management, 33*(6), 1264–1275.https://doi.org/10.1016/j.ijproman.2015.02.010

Bruton, G., & Lynn, H. (2015). Digital project management frameworks in construction. *Construction Innovation, 15*(3), 212–225. https://doi.org/10.1108/CI-01-2015-0007

Bryde, D., Broquetas, M., & Volm, J. M. (2012). The project benefits of building information modeling (BIM). *International Journal of Project Management, 31*(7), 971–980. https://doi.org/10.1016/j.ijproman.2012.03.002

Bolivar-Ramos, M. T., Garcia, F. J., & Perez, M. P. (2012). Knowledge management in construction firms: Critical success factors. *Knowledge Management Research & Practice, 10*(3), 212–223. https://doi.org/10.1057/kmrp.2012.15

Caniels, M., & Bakens, R. J. (2012). The effects of Project Management Information Systems on decision making in a multi-project environment. *International Journal of Project Management, 30*(2), 162–175.

Cao, D. (2002). Information technology in construction project management: Past, present, and future. Automation in Construction, *11*(5), 517–528. https://doi.org/10.1016/S0926-5805(02)00033-2

Cao, D., Li, H., & Wang, G. (2006). Integrating knowledge management into construction project management. *Construction Management and Economics, 24*(3), 333–343. https://doi.org/10.1080/01446190500409648

Cavalli, R. S., Fusilli, L., Pascucci, S., Pignatti, S., & Santini, F. (2008). Hyperspectral sensor data capability for retrieving complex urban land cover in comparison with multispectral data: Venice city case study (Italy). *Sensors, 8*(5), 3299–3320.

Cerutiln, M. (2004). *Information management challenges: Obstacles and hindrances to successful information management* (pp. 1–12). The Hatton.

Chang, C., Shen, L., & Hao, X. (2018). Digital transformation of construction projects: Trends and frameworks. *Automation in Construction, 93*, 123–135. https://doi.org/10.1016/j.autcon.2018.05.013

Chen, Z. (2019). Grand challenges in construction management. *Frontiers in Built Environment, 5*(31), 1–22. https://doi.org/10.3389/fbuil.2019.00031

Chen, Y., & Kamara, J. M. (2005). *The use of mobile computing in construction information management* (pp. 581–590). Association of Researchers in Construction Management.

Chen, Y., & Kamara, J. (2008). The mechanisms of information communication on construction sites. *Forum E Journal*, 1–32.

Chiou, Y., & Lee, C. (2002). Knowledge management in construction projects. *Engineering, Construction and Architectural Management, 9*(4), 339–348. https://doi.org/10.1108/eb021076

Chiou, Y., & Lee, C. (2005). Integrating information systems into project risk management. *Journal of Construction Engineering and Management, 131*(12), 1295–1305. https://doi.org/10.1061/(ASCE)0733-9364(2005)131:12(1295)

Choi, J. S. (2011). *A study on the development and application of quality control requirements for improving the quality of architectural design in open BIM environments*. Ph. D. Thesis, Gyeonggi-do, Korea: Kyung Hee University, Yongin-si.

Choo, C. W. (2002). *Information Management for Intelligent Organisation: The art of scanning the environment*. Information Today International.

Christensen, D., & Walker, D. (2004). Understanding the role of vision in project success. *Project Management Journal, 35*(3), 39–52.

Czerwinski, M., Horvitz, E., & White, S. (2004). *A diary study of task switching and interruptions* (pp. 175–182).

D4H. (2015). Incident and project management solutions for construction. https://www.d4h.com

Dahmas, S., Li, Z. L., & Liu, S. (2019). Solving the difficulties and challenges facing construction base on concurrent engineering in Yemen. *Sustainability, 11*(11), 3146–3154.

Dao, T., Nguyen, H., & Pham, L. (2017). Project information systems for large-scale construction projects. *Journal of Management in Engineering, 33*(5), 04017023. https://doi.org/10.1061/(ASCE)ME.1943-5479.0000523

Daly, J. (2021). In and out of place in federation square. *Journal of Urbanism: International Research in Placemaking and Urban Sustainability*, 1–26. https://doi.org/10.1080/17549175.2021.1950035

Davis, K. A., & Songer, A. D. (2002). *Technological change in the AEC industry: A social architecture factor model of individuals resistance* (pp. 286–291).

Dibbern, J., Goles, T., Hirschheim, R., & Jayatilaka, B. (2004). Information system outsourcing: A survey and analysis of the literature. *Database for Advances in Information Systems, 35*(4), 1–25.

Dinneen, J. D., & Frissen, I. (2020). *Mac users do it differently: The role of operating system and individual differences in file management* (pp. 1–8).

Dube, N. N. F., Aigbavboa, C. O., & Thwala, W. D. (2015). *Challenges facing construction site management in the Swaziland Construction Industry* (pp. 1–13). University Johannesburg.

Dumais, S., Cutrell, E., Cadiz, J. J., Jancke, G., Sarin, R., Robbins, D. C. (2003). *Stuff I've seen: A system for personal information retrieval and re-use* (pp. 1–8). Redmond, Microsoft Research.

Eastman, C. M. (2003). *Integration of design and construction knowledge* (pp. 206–226). Routledge.

Edology. (2014). Digital project management learning resources. https://www.edology.com

Eide, P. K. (2010). *The more things change, Acquisition reform remains the same* (pp. 1–34). U.S. Army War College, Carlisle Barracks.

Fang, E. A. H., Li, X. Y., & Lu, J. J. (2016). Effects of organizational learning on process technology and operations performance in mass customizers. *International Journal of Production Economics, 174*, 68–75.

Faris, S. (2019). *Common problems in management information systems.* [Online] Available at: http://www.chron.com/commonproblemsinmangementinformationsystems

Fitchett, S., Cockburn, A., & Gutwin, C. (2013). *Improving navigation-based file retrieval* (pp. 2329–2338). https://doi.org/10.11445/2470654.2481323. ISBN 9791450318990.

Ganesan, S., Brown, S. P., Mariadoss, B. J., & Ho, H. (2010). Buffering and amplifying effects of relationship commitment in business-to-business relationships. *Journal of Marketing, 47*(2), 361–373.

Gareev, I., & Mukhametova, D. (2018). Digital information systems in large construction enterprises. *Procedia Computer Science, 123*, 267–276. https://doi.org/10.1016/j.procs.2017.11.200

Gardiner, P. (2008). Integration of knowledge and information management in construction projects. *International Journal of Project Management, 26*(4), 411–420. https://doi.org/10.1016/j.ijproman.2007.10.006

Gelboard, J., Linder, S., & Robinson, K. (2002). Information management for construction project monitoring. *Journal of Construction Engineering and Management, 128*(4), 345–354. https://doi.org/10.1061/(ASCE)0733-9364(2002)128:4(345)

Griffith, A., & Watson, P. (2004). Effective and efficient site management. In *Construction management* (pp. 24–85). Palgrave. https://doi.org/10.1007/978-0-230-50021-1_3

Grudin, J. (2006). *Enterprise knowledge management and emerging technologies* (pp. 57a–57a).

Hall, P. (2006). *Management information systems: Managing the digital firm.* Azimuth Interactive International Publishing Services.

Hardman, K., & Marshall, J. (2000). The state and status of physical education in schools in international context. *European Physical Education Review (1356–336X0), 6*(3), 203–229.

Hartnett, E. E., & Rew, R. K. (2008). Experience with an enhanced Net CDF data model and interface for scientific data access. In *24th Conference on IIPS* (pp. 1–7).

Herb, R. (2008). Project information management in construction enterprises. *Construction Management and Economics, 26*(2), 123–132. https://doi.org/10.1080/01446190701803905

Hedberg, T., Feeney, B. A., Helu, M., & Camelio, A. J. (2007). Towards a lifecycle information framework and technology in manufacturing. *Journal of Computer Information Science Engineering, 17*(2), 1–30. https://doi.org/10.1115/1.4034132

Hendrickson, R. A. (2003). Human resource information systems: Backbone technology of contemporary human resources. *Journal of Labour Research, 24*(3), 381–394.

Herdian, R. (2014). *Printing Technology Evolution in Paper Packaging: Seminar 5*, Jakarta.

Hergunsel, M. F. (2011). *Benefits of building information modelling for construction managers and BIM based scheduling.* Worcester Polythecnic Institute.

Hodgson, J. (2001). The headache of knowledge management. *Nature Biotechnology, 19*(6), 44–46.

Hogan, P. (2009). *Top five challenges for construction project managers* (pp. 1–3). Field Wire.

Holler, J., Tsiatsis, V., Mulligan, C., Karnouskos, S., Avesand, S., & Boyle, D. (2014). From machine-to-machine to the internet of things: Introduction to a new age of internet. *Intelligence*, 1–15.

Hughes, W. P. (2008). Integrating project management information systems in construction projects. *Journal of Construction Engineering and Management, 134*(3), 175–183. https://doi.org/10.1061/(ASCE)0733-9364(2008)134:3(175)

Ilkay, M. S., & Aslan, E. (2012). The effect of the ISO 9001 quality management system on the performance of SMEs. *International Journal of Quality and Reliability Management, 29*(7), 7530–7778.

Ilyas, M. A. B., Hassan, M. K., & Ilyas, M. U. (2013). *PMIS: Boon or bane?* (pp. 1–7). IProject Management Institute.

Isikdag, U. (2012). Design patterns for BIM-based service oriented architectures. *Automation in Construction, 25*, 59–71.

Jana, K., Libena, T., & Jan, S. (2015). Support of project management methods by project management information system. *Procedia Social and Behavioural Sciences, 210*, 96–104.

Jewell, M., & Walker, D. H. T. (2005). Community of practice perspective software management tools: A UK construction company case study. In *Knowledge management in the construction industry: A socio-technical perspective* (pp. 112–129).

Jones, W. (2007). Personal information management. *The Annual Review of Information Science and Technology (ARIST), 41*, 453–504.

Jones, W., Hou, D., Sethananda, B. D., & Gemmell, J. (2010). Plans to put our digital information in its place. In *CHI' 10 Extended Abstracts on HUman Factors in Computing Systems* (pp. 2803–2812). https://doi.org/10.1145/1753846.1753866

Jones, M. (2018). Knowledge management practices in construction projects. *International Journal of Project Management, 36*(7), 987–996. https://doi.org/10.1016/j.ijproman.2018.05.006

Jones, M. (2021). Digital collaboration tools for construction project management. *Automation in Construction , 124,* 103567. https://doi.org/10.1016/j.autcon.2021.103567

Joyce, M., & Chaboyer, B. (2015). Inestigating the consistency of Stellar Revolution Models with Globular Cluster Observations via the red giant branch bump. *The Astrophysical Journal, 84*(142), 1–11.

Jung, Y., & Kim, H. A. (2016). Patterns and trends in schedule control practice of Korean general contractors. *Korean Journal of Construction Engineering and Management, 12*(4), 88–96.

Jung, Y., & Woo, S. (2004). Flexible work breakdown structure for integrated cost and schedule control. *Journal of Construction Engineering and Management, 130*(5), 616–625.

Jung, J., Su, X. M., Baeza, M., & Hong, S. K. (2008). The effect of organizational culture stemming from national culture towards quality management deployment. *The TQM Journal, 20*(6), 622–635.

Kafetzopoulos, D. P., Psomas, E. L., & Gotzamani, K. D. (2015). The impact of quality management systems on the performance of manufacturing firms. *International Journal of Quality and Reliability Management, 32*(4), 381–399.

Kalay, Y. E. (1999). *The future of CAAD: From computer-aided design to computer aided architectural design futures* (pp. 14–30).

Kaliba, C., Muya, M., & Mumba, K. (2009). Cost escalation and Schedule delays in road construction projects in Zambia. *International Journal of Project Management, 27*, 522–532.

Katranuschkov, P., Scherer, R., & Turk, Z. (2001). Intelligent services and tools for concurrent engineering? An approach towards the next gneration of collaboration platforms. In *ITcon, 6, Special Issue Information and Communication Technology in the European Construction Industry* (pp. 111–128). http://www.itcon.org/2201/10

Kaynak, H. (2003). The relationship between total quality management practices and their effects on firm performance. *Journal of Operation Management, 21*, 405–435.

Kazi, A. S., Hannus, M., Laitinen, J., & Nummelin, O. (2001). Distributed engineering in construction: Findings from the IMS GLOBEMEN project. In *ITcon, 6, Special Issue Information and Communication Technology Advances in the European Construction Industry* (pp. 129–148). http://www.itcon.org/2001/10

Kent, D., & Becerik-Gerber, B. (2010). Understanding construction project information flows. *Construction Management and Economics, 28*(5), 467–482. https://doi.org/10.1080/01446191003662368

Kermanshachi, S. (2016). Digital project management systems in construction. *Procedia Engineering*, 145, 101–108. https://doi.org/10.1016/j.proeng.2016.04.014

Kermanshachi, S., Hallowell, M., & Hadi, M. (2017). Risk management in large-scale construction projects using information systems. *Journal of Management in Engineering , 33*(5), 04017027. https://doi.org/10.1061/(ASCE)ME.1943-5479.0000510

Kleiner, D., & Buxton, B. (2008). Pledges of empire: The Ara Pacis and the donations of Rome. *American Journal of Arccheology*, 57–89.

Kokliugina, E., & Kokliugin, V. (2016). Construction project knowledge management systems. *Procedia Engineering, 153,* 112–119. https://doi.org/10.1016/j.proeng.2016.08.015

Koskela, L., & Kazi, A. S. (2003). Information technology in construction: How to realise and benefits? *Socio-technical and human cognition elements of information systems* (pp. 60–75). Idea Group Publishing.

Kudryavtseva, E., Ivanov, S., & Petrov, A. (2015). Construction project risk assessment using digital tools. *Procedia Engineering, 123*, 654–661. https://doi.org/10.1016/j.proeng.2015.10.041

Kuhne, W., & Liehr, A. W. (2009). Improving the traditional information management in natural sciences. *Data Science Journal, 8*, 18–26.

Kurtz, C. F., & Snowden, D. J. (2003). The new dynamics of strategy: Sense-making in making a complex and complicated world. *IBM System Journal, 42*(3), 462–483.

Le-Hoai, L., Lee, Y. D., & Lee, J. Y. (2008). Delay and disruption in construction projects. *Construction Management and Economics, 26*(3), 237–252. https://doi.org/10.1080/01446190701877741

Lee, S. K., & Yu, J. H. (2012). Success model of project management information system in construction. *Automation in Construction, 25*, 82–93.

Lee, D.-G., Park, Y. J., & Song, S.-H. (2018). BIM-based construction information management framework for site information management. *Advances in Civil Engineering Management, 2018*, 1–15.

Leite, F., Akins, J., Akinci, B., & Laugier, C. (2011). Visualizing construction project information using BIM. *Automation in Construction, 20*(2), 115–125. https://doi.org/10.1016/j.autcon.2010.09.005

Liao, Y., & Vemuri, R. V. (2002). Use of K-nearest neighbor classifier for intrusion detection. *Computer & Security, 21*(5), 439–448.

Liao, C., Lin, Y., & Chen, K. (2011). Integrating project knowledge management systems in construction. *Journal of Civil Engineering and Management, 17*(5), 658–667. https://doi.org/10.3846/13923730.2011.620134

Lin Y., & Su, Y. (2013). Developing mobile- and BIM-based integrated visual facility developing mobile- and BIM-based integrated visual facility. *The Scientific World Journal, 13*, 1–11. https://doi.org/10.1155/2013/124249

Liu, D., Cheng, J., & Law, K. H. (2003). Engineering information service infrastructure for ubiquitous computing. *Journal of Computing in Civil Engineering, 17*(4), 219229.

Love, P. E. D., & Irani, Z. (2003). A project management quality cost information system for the construction industry. *Information & Management, 40*, 649–661.

Mann, S., Sehgal, A., & Fung, J. (2004). *Continous lifelong capture of personal experience using eyetap* (pp. 1–21).

Marchi, M., Cesaroni, F., & Capone, G. (2018). Information management for construction project performance improvement. *Procedia Computer Science, 138*, 326–333. https://doi.org/10.1016/j.procs.2018.10.045

Marzouk, M., Azab, S., & El-Said, M. (2008). Project management systems for minimizing construction delays. *Journal of Construction Engineering and Management, 134*(8), 621–631. https://doi.org/10.1061/(ASCE)0733-9364(2008)134:8(621)

Matejka, V., & Tomek, R. (2017). Digital tools and risk management in construction projects. *Procedia Engineering, 196*, 489–498. https://doi.org/10.1016/j.proeng.2017.08.055

Maqsood, T., Finegan, A., & Walker, D. H. T. (2003). *A conceptual model for exploring knowledge channelisation from sources of innovation in construction organisations: Extending the role of knowledge management* (pp. 613–621).

Marsh, L., & Flanagan, R. (2000). Measuring the costs and benefits of information technology in construction. *Engineering Construction and Architecture, 7*(4), 423–435.

McGuinness, T., & Morgan, R. E. (2005). The effect of market and learning orientation on strategy dynamics: The contributing effect of organisational change capability. *European Journal of Marketing, 39*(11/12), 1306–1326.

Mckelvi, G., Dotsika, F., & Patrick, K. (2007). Interactive business development, capturing business knowledge and practice. *The Learning Organisation, 14*(5), 407–422.

Mehra, S., & Coleman, J. T. (2016). Implementing capabilities-based quality management and marketing strategies to improve business performance. *International Journal of Quality and Reliability Management, 33*(8), 1124–1137.

Mignerat, M., & Rivard, S. (2012). The institutionalisation of information system project management practices. *Information and Organisation, 22*, 125–153.

Miranda, A. L. (2013). *Exploring the relationship between New product development, concurrent engineering and project management to improve product development* (pp. 1–327). ProQuest LLC.

Mohamed, S. F., & Anumba, C. J. (2006). Potential for improving site management practices through knowledge management. *Construction Innovation, 6*(4), 232–249. https://doi.org/10.1108/14714170610713917

Mohamad, I., Samad, Z., & Ahmad, N. (2012). Knowledge management systems for construction SMEs. *Journal of Engineering, Project, and Production Management, 2*(2), 77–84. https://doi.org/10.32738/JEPPM.201207.0009

Newby, A. G. (2009). In building a nation few better examples can be found: Norden and the Scottish Parliament. *Scandinavian Journal of History, 34*(3), 307–329. https://doi.org/10.1080/03468750903134749

Nitithamyong, P., & Skibniewski, M. J. (2004). Web-based construction project management systems: How to make them successful. *Automation in Construction, 13*(4), 491–506.

Ofori, G. (2000). *Challenges of construction industries in developing countries: Lessons from various countries* (pp. 15–27). Gaborone.

Olsen, B. I., Lund, W. N., Ellingsen, G., & Hartvigsen, G. (2012). Document theory for the design of social-technical systems. *Journal of Documentation with Emerals Insight*, 1–30. https://doi.org/10.1108/00220411211200024347

Page, R. (2021). Digital project management for construction organizations. *Automation in Construction, 124*, 103562. https://doi.org/10.1016/j.autcon.2021.103562

PMBOK. (2004). A guide to the project management body of knowledge (3rd ed.). Project Management Institute.

PMK. (2021). Construction project management guidelines. https://www.pmk.org

Peansupap, V., & Walker, D. (2005). Exploratory factors influencing information and communication technology diffusion and adoption within Australian construction organisations: A micro analysis. *Journal of Construction Innovation, 5*(3), 135–157.

Peansupp, V., & Walker, H. (2006). Information communication technology (ICT) implementation: A construction industry perspective. *Engineering, Construction and Architectural Management, 13*(4), 364–379.

Penn, I. A., & Pennix, G. B. (2017). *Record management handbook.* Routledge. https://doi.org/10.1108/JICES-03-2021-0036

Phan, A. C., Abdallah, A. B., & Matsui, Y. (2011). Quality management practices and competitive performance: Empirical evidence from Japanese manufacturing companies. *International Journal of Production Economics, 133*(2), 518–529.

PoinfFuse. (2021). *Five challenges to digital construction.* [Online] Available at: http://www.pointfuse/fivechallengestodigitalconstruction

Prajogo, D., Huo, B. F., & Han, Z. J. (2012). The effects of different aspects of ISO 9000 implementation on key supply chain management practices and operational performance. *Supply Chain Management: An International Journal, 17*(3), 306–322.

Price, A. (2015). Construction project information systems and performance evaluation. *International Journal of Project Organisation and Management, 7*(3), 250–268. https://doi.org/10.1504/IJPOM.2015.072657

Qi, H. (2002). Construction project management information systems. *Automation in Construction, 11*(4), 435–446. https://doi.org/10.1016/S0926-5805(01)00057-0

Rakhmatullina, A. (2017). Digital knowledge management for construction projects. *Procedia Engineering, 196*, 499–507. https://doi.org/10.1016/j.proeng.2017.08.056

Raymond, L., & Bergeron, F. (2008). Project management information systems: An empirical study of their impact on project managers and project success. *International Journal of Project Management, 26*(20), 213–220.

Reddy, K. (2012). Project information management in large-scale construction projects. *Journal of Construction Engineering and Management, 138*(5), 561–569. https://doi.org/10.1061/(ASCE)CO.1943-7862.0000498

Rezgui, Y., Coper, G., Zarli, A., Samad Kazi, A., & Marache, M. (2001). *Web-based generic services for the construction virtual enterprises in the OSMOS project* (pp. 1–12).

Rezgui, Y., & Zarli, A. (2001). Editorial Information and Communication technology advances in the European Construction Industry. *Journal of Information Technology in Construction (ITcon), 6*, 83–83.

Riese, M. (2006). Proceedings of virtual prototyping worksop. In *One island East, Hong Kong: A acse study on construction virtual prototyping* (pp. 10–12).

Roe, D. (2019). *Information management remains a problem for many enterprises.* [Online] Available at: http://www.sing/cmsiswirepages/informationmanagementremainsaproblemformanyenterprises

Said, A., & El-Rayes, K. (2010). Construction project management tools and information systems. *Journal of Construction Engineering and Management, 136*(2), 234–243. https://doi.org/10.1061/(ASCE)CO.1943-7862.0000123

Salem, O., Solomon, J., Genaidy, A., & Minkarah, I. (2018). Construction project knowledge management frameworks. *Journal of Construction Engineering and Management, 144*(6), 04018043. https://doi.org/10.1061/(ASCE)CO.1943-7862.0001472

Setiawan, A. I., & Hanfan, A. (2017). Elaborating the role of network synergy capacity as a supplier's alternative terminal for achieving marketing performance. *International Journal of Business and Society, 18*(2), 245–262.

Shagiakhmetova, A., & Serdarova, A. (2018). Information technology adoption in construction management. *Procedia Engineering, 212*, 101–109. https://doi.org/10.1016/j.proeng.2018.01.013

Shane, S., Venkataraman, S., & MacMillan, I. (2009). Information management in project-based ventures. *Journal of Business Venturing, 24*(3), 271–288. https://doi.org/10.1016/j.jbusvent.2008.02.003

Shen, W., Hao, Q., Mak, H., Neelamkavil, J., Xie, H., Dickinson, J. K., Thomas, J. R., Pardasn, A., & Xue, H. (2009). Systems integration and collaborative in architecture, engineering, construction and facilities management: A review. *Advanced Engineering Informatics, 24*(2), 196–207. https://doi.org/10.1016/jaei.2009.09.001

Shenhar, A. J., Dvir, D., Levy, O., & Maltz, A. C. (2001). Project success: A multidimensional strategic concept. *Long Range Planning, 34*(6), 699–725. https://doi.org/10.1016/S0024-6301(01)00097-8

Shim, J. P., Merill, W., Courtney, J. F., Power, D. J., Ramesh, S., & Christer, C. (2002). Past, present and future of decision support technology. *Decision Support System, 33*(2), 111–126.

Siles, M. (2014). Digital project collaboration in construction. *Automation in Construction, 38*, 1–12. https://doi.org/10.1016/j.autcon.2013.10.009

Smith, P. (2014). Project cost management: Global issues and challenges. *Procedia Social and Behavioural Services, 119*, 485–494.

Stewart, R. A. (2006). *IT enhanced project information management in construction: Pathways to improved performance and strategic competitiveness* (pp. 1–25). Griffith University.

Sun, H. Y., & Ni, W. B. (2012). The impact of upstream supply and downstream demand integration on quality management and quality performance. *International Journal of Quality and Reliability Management, 29*(8), 872–890.

Sutherland, J., & Van Den Heuvel, W. J. (2002). Enterprise application integeration and complex adaptive systems. *Communication of the ACM, 45*(10), 59–64.

Sydorenko, O. (2021). Information management for construction project monitoring. *Procedia Computer Science, 194*, 295–302. https://doi.org/10.1016/j.procs.2021.10.036

Tambwe, O. T., Aigbavboa, C. O., Akinradewo, O. I., & Adekunle, P. A. (2025). Measures to address cyber-attacks in construction project data management processes: A cybersecurity perspective. *IET Information Security, 2025*(1). https://doi.org/10.1049/ise2/7398742

Teevan, J., Alvarado, C., Ackerman, M. S., & Karger, D. R. (2004). *The perfect search engine is not enough: A study of orienteering behaviour in directed search* (pp. 415–422). ACM Digital Library. https://doi.org/10.1145/9859692.985745.S2CID1180143

Tempini, N. (2017). Digital data and construction project performance. *Automation in Construction, 81*, 23–36. https://doi.org/10.1016/j.autcon.2017.03.008

Tombesi, P., Gibello, L., Milan, L., & Chiorino, C. (2008). Se le Archistar Sbagliano, Special Report. *Il Giornale dell'Architettura, 7*(65), 4–9.

Tombesi, P. (2004). Sloppily built a precisely lose? The technology of the curtain and the ideology of Disney Hall. *Architectural Research Quarterly, 8*(3–4), 247–259.

Tombesi, P. (2009). On the cultural sepration od design labour. In *Building the future: Recasting labour in architecture* (pp. 117–136). Princeton University Press.

Tombesi, P., & Whyte, J. (2015). Risk-based project information systems in construction. *Construction Management and Economics, 33*(2), 107–121. https://doi.org/10.1080/01446193.2015.1011234

Toyota, Y., Oguchi, K., & Nagata, M. (2004). Acoustical design of Walt Disney Concert Hall. In *Proceedings of ICA* (Vol. 2004, pp. 2943–2946).

TrueLook. (2019). *Five project management challenges in the construction industry*. [Online]. Available at: http://www.truelook/fiveprojectmanagementchallengesintheconstructionindustry

Trentin, G., Repetto, P., & Corti, A. (2012). Digital learning and information management in construction. *Procedia–Social and Behavioral Sciences, 69*, 1534–1542. https://doi.org/10.1016/j.sbspro.2012.12.088

Uusitalo, V., Haapasalo, H., & Aaltonen, K. (2019). Information and knowledge management for construction project risk. *International Journal of Project Management, 37*(7), 907–921. https://doi.org/10.1016/j.ijproman.2019.05.007

Vallas, S. P., & Christin, A. (2018). Work and identity in an era of precarious employment: How workers respond to "personal branding" discourse. *Work and Occupations, 45*(1), 3–37.
Van Den Heuvel, W. J., & Mammar, Z. (2003). Moving towards a framework to compose intelligent web services. *Communication of the ACM, 46*(10), 103–109.
Vereecke, A., & Muylle, S. (2005). Performance improvement through supply chain collaboration: Conventional wisdom versus empirical findings. *Vlerick Leuven Gent Working Paper Series* (pp. 1–38).
Vitt, E., Luckevich, M., & Misner, S. (2002). *Business intelligence: Making better decisions faster*. Microsoft Press. http://proquest.safaribooksonline.com/9780735626607.
Vuori, V., & Okkonen, J. (2012). Refining information and knowledge by social media applications: Adding value by insight. *The Journal of Information and Knowledge Management System, 42*(1), 117–128.
Walasek, P., & Barszcz, D. (2017). Construction project information management systems: A review. *Procedia Engineering, 196*, 515–523. https://doi.org/10.1016/j.proeng.2017.08.060
Wang, S. (2003). Project management information systems for large construction projects. *Automation in Construction, 12*(6), 735–742. https://doi.org/10.1016/S0926-5805(03)00067-1
Wang, X., & Huang, J. (2006). The relationships between key stakeholders' project performance and project success: Perceptions of Chinese construction supervising engineers. *International Journal of Project Management, 24*, 253–260.
Warkentin, M., Johnston, A., & Shropshire, J. (2000). Management information systems in construction projects. *Information & Management, 37*(5), 263–272. https://doi.org/10.1016/S0378-7206(99)00058-2
Wender, K., & Hubler, R. (2009). Enabling more human-oriented exploration of complex building information models. *Journal of Computing in Civil Engineering, 23*(2), 84–90.
Westerveld, E. (2003). The project excellence model: Linking success criteria and critical success factors. *International Journal of Project Management, 21*(6), 411–418. https://doi.org/10.1016/S0263-7863(03)00004-4
Whittaker, S. (2011). Personal information management: From information consumption to curation. *Annual Review of Information Science and Technology, 45*(1), 1–62. https://doi.org/10.1002/aris.2011.1440450108.ISSN1550:8382
Whittaker, S., Mattews, T., Cerrut, J., Badeness, H., & Tang, J. (2011). *Am I wasting my time organising email? A study of email refinding* (pp. 3449–3458). Computer Science Bibliography.
Wickforss, A., & Lofgren, M. (2007). Knowledge sharing and information systems in construction. *International Journal of Project Organisation and Management, 2*(3), 202–214. https://doi.org/10.1504/IJPOM.2007.017451
Wilson, I., Harvey, S., Vankeisbelck, R., & Kazi, A. S., 2001. Enabling the construction virtual enterprise: The OSMOS approach. In *Itcon, 6, Special Issue Information and Communication Technology Advances in the European Construction Industry* (pp. 83–110).
Wong, C. T., & Teoh, S. H. (2008). Construction project information management: Lessons from practice. *Journal of Construction Engineering and Management, 134*(7), 492–500. https://doi.org/10.1061/(ASCE)0733-9364(2008)134:7(492)
Yeo, K., & Ning, C. (2006). Construction project information management for small and medium-sized enterprises. *Construction Management and Economics, 24*(10), 1075–1085. https://doi.org/10.1080/01446190600950407
Yen, H. R., Li, Y. E., & Niehoff, B. P. (2008). Do organisational citizenship behaviours lead to information system success? *Information & Management, 45*, 394–404.
Yoon, S., Choi, H., & Park, M. (2003). Digital project management and information sharing in construction. *Journal of Construction Engineering and Management, 129*(3), 290–298. https://doi.org/10.1061/(ASCE)0733-9364(2003)129:3(290)
Yu, X., Shen, Q., & Li, H. (2005). Construction project knowledge management systems: Practices and challenges. *Automation in Construction, 14*(6), 727–739. https://doi.org/10.1016/j.autcon.2005.04.003

Yunis, M., Jung, J., & Chen, S. M. (2013). TQM, strategy, and performance: A firm-level analysis. *International Journal of Quality and Reliability Management, 30*(6), 690–714.

Zhang, S., Teizer, J., Lee, J., Eastman, C. M., & Venugopal, M. (2013). Building information modelling and safety: Automatic safety checking of construction models and schedules. *Automation in Construction, 29*, 183–195.

Zhiliang, M., & Liang, Q. (2004). *A framework of information management system for construction projects* (pp. 1–8). Tsinghua University.

Zika Viktorsson, A. (2006). Knowledge management in construction projects. Engineering, *Construction and Architectural Management, 13*(2), 143–153. https://doi.org/10.1108/09699980610659850

Chapter 4
Fourth Industrial Revolution Tools in Construction

Abstract This chapter gives an overview of the fourth industrial revolution in the construction industry. Specific attention is given to fourth industrial revolution tools that can be used to improve construction information management.

4.1 Introduction

Information is a vital asset in the construction industry, as it underpins decision-making, resource management, and project delivery (Kubjana et al., 2024). Traditional methods of handling construction information have often been plagued by inefficiencies, fragmentation, and poor collaboration among stakeholders, resulting in delays, cost overruns, and disputes. To overcome these challenges, the industry is increasingly turning to innovative solutions enabled by the Fourth Industrial Revolution (4IR) (Adekunle et al., 2022). These advancements have redefined the way information is generated, stored, shared, and utilised, thereby improving project efficiency and transparency. The integration of mobile devices, cloud-based platforms, and smart wearables has allowed construction professionals to access and update information in real time, regardless of location. Likewise, building information modelling (BIM) has emerged as a critical tool for digital collaboration, offering a unified platform where project data can be visualised and analysed across the project lifecycle (Otasowie et al., 2023). More advanced technologies, such as artificial intelligence, robotics, augmented reality, and big data analytics, provide predictive capabilities and automate repetitive tasks, reducing human error and enhancing efficiency. Emerging frontiers, including blockchain-based BIM, quantum computing, and 5G networks, hold transformative potential for securing data integrity, enabling ultra-fast communication, and managing increasingly complex construction systems.

© The Author(s), under exclusive license to Springer Nature Switzerland AG 2026

P. Adekunle et al., *Emerging Trends in Construction Information Management*,
https://doi.org/10.1007/978-3-032-09679-1_4

4.2 Overview of the Fourth Industrial Revolution

Information Technology (IT), also known as registration and correspondence innovation, has radically altered the way people live, study, work, and play (Froese, 2010). Many development organizations do not appear to like the beneficial changes and benefits that the new innovation was bringing to organizations in many sectors of the economy (Mak, 2000). PCs were only used in a few construction firms in the 1980s, but most construction firms used PC technology for a significant portion of their core operations, such as bookkeeping and payroll (Froese, 2010). Few of them have built appropriate approaches or techniques for utilizing data innovation.

By the end of the 1980s, about eight years after the introduction of solid PC hardware, a few companies had reached a point where their employees on a significant number of their larger ventures were experiencing the benefits of the new technology through the use of arranging, drawing, accounting page, and word-processing programming bundles (LCI, 2008: 1). At each stage of the construction cycle, a variety of product bundles are now available to all disciplines of construction professionals (Levitt, 2008: 1). They aid with a wide range of tasks, including building representation, plan evaluation, project management, data storage and recovery, cost assessment, primary investigation, site administration, and office management, among others.

4.3 Mobile Device Management

Mobile device management (MDM) is the most frequent method of improving corporate information security by inspecting, monitoring, and obtaining mobile devices such as desktops, phones, and tablets that are used in projects (Abowd & Mynatt, 2000). IT groups and administrators can use mobile device management solutions to govern and distribute security protocols to cell phones that access sensitive corporate data, ensuring the security of the company (Saha & Mukherjee, 2003). With an ever-increasing number of employees utilizing one or more of these devices, businesses of all shapes and sizes are now opting to cell phone the board for improved data security and expanded functionality (Oloufa et al., 2003). MDM systems allow IT managers to create enterprise-grade security strategies for mobile devices, allowing them to be used in the workplace (Aziz et al., 2004).

IT administrators can use mobile device management (MDM) to safely monitor and manage the cell phones that access sensitive company data (Liu et al., 2003: 220). It entails storing basic information about cell phones, selecting which programs can be installed on the devices, tracking down and retrieving lost or stolen devices. Mobile device management (MDM) arrangements have evolved into Enterprise Mobility Management (EMM) systems as cell phone reception has grown (Tenopir, 2003: 14).

Cell phones now have more capabilities than at any previous moment in history, prompting many businesses to embrace a mobile or flexible first workforce (Gyampoh-Vidogah & Moreton, 2003: 158). Individual and business-owned cell phones are the primary means of accessing or interacting with corporate information in these circumstances (Satyanarayanan, 2001: 12). Many firms use third-party mobile device management (MDM) software like Mobile Device Manager Plus to handle cell phones (Rebolj et al., 2001: 5).

4.4 Cloud Management

Cloud computing is a new way of thinking that is gaining traction and acceptance (Armbrust et al., 2009: 28). Cloud providers use shared servers to provide processing, storage, and programming services to consumers (Kozuch et al., 2009: 3). Both Gartner and IDC have rated the reception of distributed computing as solid (IBM Cloud, 2021: 1). Cloud administrations include, among other things, figure on-demand, online capacity, on-line/shared office, key-value store management, and email (Microsoft Cloud, 2021: 1). Amazon AWS, GoGri, and Rackspace are examples of public cloud providers. HP, Google, IBM, and Microsoft are among the companies that have contributed to the cloud (Moreno-Vozmediano et al., 2009: 4). Moreover, other open-source projects such as Eucalyptus and Open Nebula, as well as various examination projects such as Tashi, Reservoir, and Open Cirrus, are in the works. Virtualization, administration-assisted registering, and utility processing have all aided distributed computing (IDC, 2019: 1). There are a few requirements for successful distributed computing. These include low costs, consistent SLAs, security guarantees, high accessibility, energy productivity, and accurate bookkeeping (GoGrid, 2012: 2). Successful management of cloud assets and services is the key to satisfying these requirements. From raise to provisioning, booking, checking, disappointment management, and closure, this covers every aspect of the server project lifecycle (Bhargav-Spantzel et al., 2006: 272).

As seen in Fig. 4.1, mists are planned as a layered design. Infrastructure as a Service (IaaS), Platform as a Service (PaaS), and Software as a Service (SaaS) layers are included in cloud engineering (Forell et al., 2011: 881). IaaS refers to the calculating and capacity foundation provided as a service, PaaS refers to the staging and moreover arranging stack provided as a service, and SaaS refers to the delivery of programming as a service over the Internet. Private, public, or half-breed mists can be used to transmit such cloud designs. Private mists are created in-house inside projects, while public mists are purchased online. As seen in Fig. 4.1, mixture mists combine public and private mists (Forell et al., 2011: 881). Effective asset management at the various layers of the stack, across several measurements of framework activity execution (SLA consistency), strength, and power, is a critical requirement in this big number of firms (Bash et al., 2011: 4). Provisioning, booking, checking, disappointing executives, execution enhancement, and effective power supply are all included. Customary checks and executive advancements were designed for large

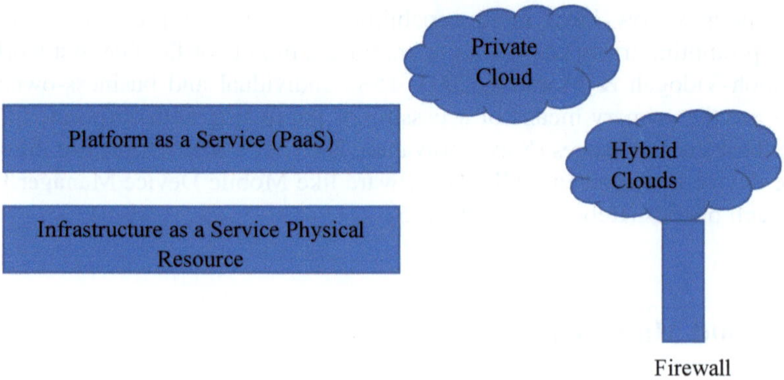

Fig. 4.1 Cloud architecture. *Source* Forell et al. (2011: 882)

company circumstances and are frequently required for single client conditions sent across thousands of servers (Wang et al., 2011: 388). Cloud frameworks assist customers deal with multi-occupant situations by providing processing frameworks and apps. These frameworks may scale up to a hundred thousand servers and serve a massive number of users (Avetisyan et al., 2010: 43).

4.5 Smart Wearables

The term smart has no universally accepted definition among researchers, and it is defined in a variety of ways by different analysts (Adapa et al., 2017: 340). Smart objects, for example, are free advanced products that are upgraded by sensors, processors, and organization offices (Kortuem et al., 2010: 45). Brilliant objects move a few chunks of usage logic that allow them to recognize their surroundings and interact with people (Garcia et al., 2018: 3). Brilliant wearables and shrewd clothes are both covered by wearable processing. Smart wearables are devices with remote sensors that are embedded in articles of clothing or accessories and can be worn by customers all day (Cheng & Mitomo, 2017: 529). Installed PCs in anything that people normally employ to cover or beautify their bodies are also known as wearable registering (Posland, 2011: 6). It's a wearable electronic device that provides specific forms of help to clients.

However, there is no one-size-fits-all approach to smart wearables. Wearables are divided into five categories by Global Data Corporation (IDC, 2017: 1): watches, wristbands, clothes, earwear, and others. Dimou et al. (2017: 71) classified smart wearables into six categories: entertainment, way of life, wellness, clinical, modern, and gaming. Wearables for mobility, Augmented Reality (AR), control devices, and clever gloves are all part of the fun. Wearables are used for video and voice calling, as well as signal control. Wearables for measuring distance traveled, pulse, and internal heat level are included in the wellness category. Clinical wearables include amplifiers, cardiovascular monitoring, patient remote monitoring, and bionics. Wearables used in remote and hands-free activities related to modern and business goals are included in contemporary. Wearables that use augmented reality for gaming are included in gaming. Wearables are also divided into three categories: accessory or bracelet type, watch type, and head-mounted display type (Yang et al., 2016: 256). Wearable watches might get cell phone, instant message, and email notifications without having to pull the phone out of the pocket. Jewelry or wristband-style wearables are mostly used to monitor people's health over time, whereas head-mount show-style wearables are suitable for Virtual Reality (VR) content and three-dimensional (3D) gaming.

4.6 Building Information Modelling

Building Information Modeling or Building Information Management is abbreviated as BIM. Planners, engineers, land engineers, project workers, makers, and other development specialists can plan, plan, and build a design or work inside one 3D model (Motamedi & Amin, 2009: 238). It can also extend to the activity and board of structures utilizing data that structure or design owners approach (Masood et al., 2014: 1). Even after the structure is built, this information allows legislators, regions, and property managers to make informed decisions based on data obtained from the model (Charef et al., 2011: 189). Previously, data concerning a given structure plan was communicated through outlines and sketches. It was difficult to visualize elements and prerequisites using this 2D methodology (Arayici et al., 2011: 7). Then came CAD (Computer Aided Design), which made it easier for drafters to see the benefits of plans in a more advanced climate (Azhar, 2011: 241). Later, CAD evolved into 3D, bringing more realistic graphics to outlines. BIM (Building Information Modeling) is the current standard; yet, it is much more than a 3D model (Kouider et al., 2017: 57). BIM objects, or the components that make up a BIM model, are smart, calculate, and store data (Terreno et al., 2015: 83). When a component is modified, BIM programming updates the model to reflect the change. This allows the model to remain stable and composed during the interaction, allowing underlying specialists, planners, MEP engineers, fashioners, project directors, and project employees to collaborate more effectively. BIM, in general, refers to the collaboration and sharing of information among all parties involved in the development and lifecycle management of constructed resources (Barlish & Sullivan, 2012: 153). Nonetheless, the true power of BIM is contained in the "I" (data) (Love et al., 2016: 157). The information gathered from start to finish isn't just saved; it's important.

4.7 Artificial Intelligence

For a long time, the overall development industry's usefulness yield has been improving only slightly or not at all (Abdelgawad & Fayek, 2010: 1029). Perhaps the biggest explanation is found on the construction site itself (its strategic failures). To begin, roughly 33% of the time spent at construction sites is spent on transportation, changes, vacations, and the search for resources (Abotaleb et al., 2016: 16). Second, development initiatives are frequently confronted with issues arising from development delays (Guo et al., 2010: 2). Around 90% of megaprojects are completely postponed, are beyond budget, or show other variations from the original concept (Du et al., 2020: 3). These characteristics result in unnecessary costs for organizations involved in development cooperation (Fiorce et al., 2016: 412).

Simulated intelligence-based arrangements can help in a variety of ways during the development cycle. Man-made intelligence can help to improve development execution planning, development successions, and project management executives while keeping all partners up to date (Kim & Cho, 2019: 3). Furthermore, AI can improve the usefulness of development execution itself. By comparing the sophisticated twin of the structure available in the Building Information Modeling (BIM) cloud to the real actual portrayal, artificial intelligence-based devices could identify probable disputes, deferrals, and alterations within the development interaction (Mena et al., 2014: 143). For consecutive image recording and 360° laser examinations, robots and robots might already be used (Pan et al., 2020: 702). These images and estimates provide the necessary data for the AI to track the ideal progress of the actual construction against the plan stored in the BIM cloud (Son et al., 2019: 28).

Furthermore, comparing real construction progress to the digital twin allows for programmed identifiable proof of errors/deviations from the original structure design during the development stage (Perez-Ramirez et al., 2019: 604). This allows us to respond swiftly and, as a result, conduct preliminary investigation. This avoids the need for more expensive treatment in the future (for example in the event that lines or links have been introduced in some unacceptable area, they can be moved before all dividers are shut). Potential deferrals and alterations may prompt the AI to carry out a pre-programmed change in the BIM cloud-stored construction plan (Pan et al., 2020: 417). All parties, such as distributors and producers, might be educated as a result of this principle. Since a result, construction site coordination would be eased, as sub-project workers and providers could adjust their expectations to the new schedule (Selakov et al., 2014: 81). As a result, an AI-driven construction cycle could be the major enabling impact for on-time delivery of construction supplies (Nowakowski et al., 2017: 108).

Mistakes are a major cause of construction delays and expense overruns. When you look at statistics from countries like Germany (6000 mishaps each year) or Japan (300 deaths and 15,000 injuries at construction sites on a regular basis), the importance of increased security at construction sites becomes clear (Wang et al., 2017: 68). The use of AI as PC vision in conjunction with cameras on the construction site could drastically reduce the number of accidents (Wei et al., 2020: 3). The potential for this can already be seen in some construction sites, as many organizations have begun offering image recognition algorithms that can help construction workers discover equipment more quickly (Zhang et al., 2019: 992).

When a construction worker need a specific device, such as a hot saw, they just ask the AI by typing "Where a roundabout is saw not being used?" on their phone. The AI then analyses the images from the multiple cameras on site and displays the location of the nearest hot saw (Chokor et al., 2016: 1589). The development laborer then requests that the AI send approved personnel up to their area with the hot saw, which the AI complies (Gogna & Tayal, 2013: 504). Simultaneously, AI improves both device usage and efficiency, since construction workers no longer have to walk about the construction site for hours seeking for instruments and equipment (Sharif et al., 2017: 894). The advantage of AI over sensors and IoT is that it is far more versatile. When the cameras are installed, each item can be quickly assessed to see if it is IoT enabled (Zhou et al., 2015: 338). True models demonstrate the possibility for a 40% increase in labor usefulness, as well as project completion at over 10% less cost, as the hour of construction laborers and hardware may be utilized more effectively (Hosseini et al., 2019: 369).

Another advantage of the AI-based system over sensors/IoT is that it can be programmed to alert employees whenever it detects potential security threats (for example, when a gadget is not properly stored or workers are not wearing security gear) (Deng et al., 2020: 374). A capable individual would be dispatched to the region to eliminate the security threat. Aside from accident avoidance, AI can also assist administrators of heavy and portable equipment at construction sites (Kanan et al., 2018: 74). Administrators frequently carry out several tasks at once while operating a system (Han et al., 2018: 2682). This leads to a high rate of errors and poor performance (for example a tractor loses material from its digging tool prior to being put at the spot it is intended to be). Man-made intelligence has the potential to dramatically increase the use of huge equipment by assisting the machine administrator with regular tasks (Salhi et al., 2019: 2). These tiresome tasks are ideal for AI-based devices to take over, as they can quickly learn and adapt to these tasks. The administrator can then focus on more complicated and worthwhile projects (Bock, 2015: 114). AI is also used to improve the security of working construction equipment. Continuously monitoring the weather, checking for potential repercussions with humans or protests, and alerting the administrator before catastrophes or stopping the machine (Ding et al., 2018: 119). Simulated intelligence allows administrators to operate faster and more accurately for extended periods of time. According to a company that has gained actual experience in the control and robotization of gear, AI might increase the efficiency of portable hardware by up to 60% (Arashpour et al., 2016: 263).

4.8 Automation and Robotics

Advanced mechanics and mechanized frameworks have the potential to disrupt and enhance the development industry as well as the Architecture, Engineering, and Construction (AEC) sector as a whole (Ajoudani et al., 2018: 956). Development is a labor-intensive field. Mechanical frameworks and robotization have proven to be quite tempting in various fields for lowering labor costs while increasing efficiency and quality (David, 2015: 4). Furthermore, automated frameworks can reduce wounds and relieve specialists from dangerous tasks. Ordinary development tactics have reached their limits, and improvements in mechanization and mechanical technology may be able to overcome the construction industry's efficiency issues (Gupta et al., 2019: 135). Mechanical technology frameworks for construction were produced in the 1960s and 1970s at the same time as several ventures, such as the automobile industry, began their computerization; yet, mechanical technology's reception in the development industry has been extremely slow (Rakha & Gorodestky, 2018: 253). It should be noted that different initiatives have different levels of computerization in development (Kumar et al., 2008: 2). While organizations believe that mechanical technology and computerized frameworks will improve usefulness, well-being, and security, a study of 11 large development organizations and government offices in Europe found that there are significant risks to reception, including significant execution costs and high business and specialized risks (Pan et al., 2018: 83). It also acknowledges that the construction industry has been slow to adopt mechanical technology and new innovations in general; however, it contends that mechanization and mechanical technology are influencing the industry in general, and that they will play a role in future business advancements (Stilger et al., 2017: 90). Other mechanical developments, such as the Industry 4.0 worldview, Building Information Modeling (BIM), detecting innovations, and computerized reasoning, may help advanced mechanics gain traction in the construction industry (Sauder et al., 2017: 3).

As shown in Table 4.1 (Delgado et al., 2019: 3), there are four general classes of mechanization and mechanical developments for construction: Drones and autonomous vehicles, as well as exoskeletons, are all examples of off-site building frameworks. The first development robots were designed in Japan to broaden the scope of creating components for specific residences. (Off-site construction frameworks, Classification 1) Because of the effective employment of robots in Japan's vehicle assembly industry, these robots have been well received. Following that, development robots began to appear on construction sites, and robotized building destination frameworks were developed (Category 2: in the vicinity mechanized and automated frameworks). Finally, exoskeletons are wearable mechanical devices that boost the client's capabilities. Exoskeletons are not strictly a mechanical framework because they enhance the laborer's abilities rather than completely replacing them (Bock, 2015: 115). However, exoskeletons were included because this review focuses on all equipment developments that improve development workouts (Castro-Lacouture, 2009: 1068). Similarly, this distinction will become less visible as time

Table 4.1 Examples of robotic, automated and autonomous systems

Category	Description, types and use-cases
Off-site automated prefabrication system	Production of building components at off-site locations in an automated manner
	• Building component manufacturing (BCM)
	• Large-scale prefabrication (LSP)
	• Additive manufacturing (3D printing)
On-site automated and robotic systems	Automated and robotic systems used directly on the construction site
	• Single task construction robots (STCRs) for bricklaying, steel-support get together, steel welding, divider painting and so forth
	• Robotic on location plants • Swarms and robots for building part gathering
Drones and autonomous vehicles	Terrestrial, aerial or nautical vehicles that can be piloted remotely or which are autonomous
	• Access to outrageous and hazardous conditions
	• Surveying, investigation and observing • Automated penetrating, uncovering and earthmoving
Exoskeletons	Wearable devices that collaborate with the customer rather than a robot that completes the task on its own
	Improve the utility of laborers by allowing them to lift heavy loads, reduce fatigue, and work with equipment in unusual positions
	• Reduce wounds

Source: Delgado et al. (2019: 3)

passes. Exoskeletons, for example, necessitate a high level of computerization, and a significant amount of human–robot collaboration is likely to exist (Goessens et al., 2018: 459). In this sense, before construction sites were fully devoid of human specialists, it was likely that robots, mechanized frameworks, and more laborers would perform in unison (Czarnowski et al., 2018: 180).

4.9 Augmented Reality

Augmented reality (AR) is a live, reproduced perspective on a physical, true climate whose components are raised (or enhanced) by PC created tangible info. It is one of the three extended reality (XR) virtual conditions, which also includes virtual reality (VR) and mixed reality (MR) (Yoders, 2021: 1). Computer produced reality replaces the current reality with a replica, blended reality combines the current reality with a more advanced form, and enhanced reality takes the current reality and adds to it, such as a 3D model of your idea for engineering (Schall et al., 2013: 1534). The data about the client's surrounding genuine world becomes intelligent and can be carefully controlled with the help of cutting edge expanded reality innovation like PC vision and article recognition. PC code should infer correct arrangements in extended reality, independent of the camera or camera images (Schoenfelder & Schmalstieg, 2008: 84).

Using cell phones and 3D models, a 3D model of a suggested design can be projected onto an existing space in development and engineering projects (Klinker et al., 2001: 398). For a long time, augmented reality (AR) has been used in video gaming and media entertainment to display a real-world image that connects to one created using PC designs (Pentenrieder et al., 2007: 33). Its popularity grew in the engineering and construction industries when project managers, like as those at Seattle's BNBuilders, began using Apple iPads and other mobile devices on construction sites to display customers proposed plans in relation to actual conditions (Golparvar-Fard et al., 2009: 131).

Seeing an Autodesk Revit or other 3D model in action makes space planning and representation much easier (Schall et al., 2010: 5). AR was previously restricted to large design, design, and development firms with large innovation teams that could spend hours integrating Revit models with natively built 3D-game-motor models, but the technology has now been democratized and is available on a project-by-project basis, allowing small businesses and even sole proprietors to take advantage of it (Shin & Dunston, 2008: 883). Increased reality has a lot of plans, and development also uses past perception. It could be used for plan inquiry to identify conflicts by just walking through your completed model (Coté et al., 2012: 1). It has all of the qualities needed for a constructability audit, allowing the designer and project manager to collaborate on modifications that need to be made to the plan and development due to constructability difficulties. It can even aid in the fabrication of building components (Columbia et al., 2008: 1203).

4.10 Innovative Materials

Scholarly and modern initiatives all over the world are continually drawn in to foster new smart materials that can provide effective alternatives to conventional construction materials, improve energy efficiency in structures, or revamp, repair, and ensure existing foundations (Yudenko et al., 2019: 3). Before a true market entry, this new era of materials should be examined, approved, field tested, and possibly presented to raise expectations about their long-term conduct and execution (Sukharev, 2014: 25). In this regard, the significant contributions in the Special Issue on Innovative Materials for Construction provide an assortment of unique examination and recent fads in the field of inventive materials and advancements proposed for the construction industry, with a special focus on manageable materials for the structure business of the future, while keeping in mind creative arrangements appropriate for old developments and social legacy (Yudenko et al., 2019: 3).

The determination of the best-performing materials ready to meet the undertaking target is the most important phase of any development configuration (Lukustsova et al., 2019: 132). Despite the fact that this is an important task in the planning of a construction project, because the mechanical, practical, and actual properties of construction materials have a significant impact on the overall project execution, there are currently no systematic strategies to guide architects in this decision (Solovyova et al., 2019: 64). A novel methodology based on the Analytic Hierarchy Process (AHP) was presented to remedy this gap (Fedoseev et al., 2018: 33). The designers accepted the AHP technique in the design of a composite framework structure management, indicating the practicality of the suggested model capable of determining the best material mix and ensuring the best exhibition (Lee et al., 2020: 1738).

Shrinkage that occurs during the cement fix addresses a big difficulty in substantial innovation, impeding fitting execution and limiting the strength of substantial designs (Masanaga et al., 2018: 3018). To address this issue, a comparison was made between the qualities and execution of another system research and application (SRA) and a conventional SRA when applied to a considerable blend based on Portland concrete. In terms of further increased cement sturdiness against freeze-defrost cycles, the new shrinkage-diminishing specialist's superiority was confirmed. In addition, the review's authors examined the instrument by which the planned SRA may reduce concrete shrinkage (Frigione et al., 2019: 3502). The consolidation of modern side-effects in development materials is a reasonable and helpful manner to take advantage of waste to create new materials, dispensing with the issue of waste treatment and, simultaneously, keeping away from the utilization and exhaustion of new regular assets (Sarcinella et al., 2020: 2055).

In this regard, it was suggested that scoop heater slag be combined with the bitumen emulsion production process in recovered black-top asphalts that had been set aside for reuse (Cho et al., 2020: 4993). The researchers demonstrated that by properly determining the black-top blend components (RAP, scoop heater slag, water, and emulsion), street asphalts with excellent mechanical qualities and toughness can be produced. Two distributions arose from an Italian–logical coordinated effort to advance inside mortars in view of diverse folios (ethereal and pressure driven lime, gypsum, and concrete) holding a unique viable phase change material (PCM) ready to further expand the energy effectiveness of structures (Al-Imran et al., 2020: 4198). Since PCMs may change their physical state in response to changes in ambient temperature, the use of a suitable PCM in mortars can reduce indoor temperature variations, resulting in improved human comfort and lower enthusiastic use (Madej, 2020: 2366). The use of non-harmful, low natural effect materials and waste materials/side-effects from various businesses were used to determine the parts that would deliver a green composite PCM framework (Chin et al., 2020: 2366). The authors of this study demonstrated that, depending on the type of folio, it was possible to generate mortars with adequate mechanical properties by improving their arrangement (Vesova, 2019: 76).

As a result, the warm execution estimates of the best-performing mortars, namely those based on concrete and gypsum, confirmed that the expansion of the tested PCM in mortars causes a reduction in the maximum attainable temperatures during the hot season (Acs & Estrin, 2018: 502). It also causes an increase in base temperatures during the viral season, resulting in lower heating and cooling needs, demonstrating the new PCM material's ability to achieve energy reserve funds (Sukharev, 2018: 15).

4.11 Autonomous Vehicle

The use of autonomous vehicle and hardware could present a significant opportunity for the construction industry to improve productivity (Thrun et al., 2006: 663). Independent vehicles are currently mostly used in mining and farming operations (Kahneman & Frederick, 2002: 51). For example, Komatsu has extracted multiple billion tons of surface material on its own, and it plans to send 150 additional trucks to the Canadian oil sands over the next seven years (Savoie & Bedeau, 2019: 1). Caterpillar, not to be outdone, is sending an army of autonomous trucks and shot drills to the Rio Tinto Koodaideri iron mine in Western Australia. Backhoes, dozers, and automated bricklayers are examples of independent, semi-independent, and controller hardware that are currently being developed in the construction industry (Zhao & Thorpe, 2000: 150). Independent Vehicles (AVs) run errands in dangerous or remote locations, in fields where fixed robots are restricted, and are capable of advancing to the ideal location to perform specific tasks, such as merchandise transportation, investigation, and apparatus control (Tran & Ha, 2004: 3). The variety of uses in which these vehicles should work delivers an assortment of vehicle arrangements. Notwithstanding AVs' incredible potential (i.e., wellbeing and usefulness), their utilization isn't far reaching because of the intricacy to characterize the automated framework arrangement (Zanasi et al., 2001: 417).

There is evidence that using independent vehicles and hardware improves efficiency, resulting in increased productivity (Setlur et al., 2003: 103). For example, Rio Tinto announced in 2017 that driverless trucks were more successful than their monitored counterparts in mining jobs, as the autonomous vehicles operated for 700 h on average and had 15% cheaper unit costs (Bedeau & Savoie, 2019: 1). Rio Tinto estimates that their independent truck armada will increase efficiency by billions of dollars. Apart from increased utility, independent vehicles and hardware may be able to help tackle present work shortages in the construction industry, as well as improve security on construction sites (Ha et al., 2004: 134).

4.12 5G Network

Despite the fact that 5G represents a significant advancement over 4G and WIFI, an unexpectedly large improvement is always required for end users to accept new technology (Xiang et al., 2020: 1). Currently, the great majority believe that IoT innovations will enhance the portable development machine business with capabilities such as predictive support, data examination, and representation and warning (Ghost et al., 2010: 10). In any case, we think they're fantastic must-have advancements. Because 5G may have a large number of small base stations, each of which consumes a significant amount of energy, the value provided by predictive assistance is unlikely to compensate for the additional cost of 5G (Roh et al., 2014: 107). Getting reinforcement cars ready can often be a more successful and cost-effective solution (Ansari et al., 2018: 3971). The harsh atmosphere on the working site, such as residue obstruction and goliath equipment, will also exacerbate mmWave's insufficiency (Letaief et al., 2019: 85).

As a result, we accept more practical scenarios such as controllers and self-working portable equipment since 5G allows us to achieve things we couldn't do before (Saad et al., 2020: 135). The working climate in various dangerous traditional endeavors, such as remote support of subterranean pipelines, remote avalanche rescue, underground mine digging, and so on, is risky and harmful to the human body (Kim et al., 2012: 272). Although controllers can be used with a wired organization for today's projects, the adaptability is limited by the vehicle's link, thus controllers are only used in certain situations (Han et al., 2015: 187). Because 5G allows controllers to operate without being constrained by connectivity, controller usage is increased. In this case, cameras are usually installed on the machines to collect general climate data (Yoshida et al., 2019: 3). Because they frequently require several cameras to obtain data, and WIFI communication speeds are limited, they are unable to add more cameras to create the depth data, resulting in decreased utility even with the finest administrators (Dadhich et al., 2018: 179).

Since 5G will embrace virtual reality innovation, the controller's burden will be significantly reduced (Gupta et al., 2015: 1207). 5G safeguards the controller's productivity and accuracy better than earlier organization advances. Self-working devices are another anticipated application. The image can also be processed in the local cloud, assisting profound learning-based picture handling models (Xiang et al., 2020: 1). The order would then be able to be despatched directly from the machines. To save money, many academics point out that instead of introducing more equipment, we can use a cell phone as a mediator to transfer data (Xiang et al., 2020: 2).

4.13 Quantum Computing

Quantum computing is poised to disrupt virtually every industry. These superfast machines allow any industry that relies on finding out a way to deal with problems that are beyond the capabilities of even today's supercomputers to work faster, progress cycles, and improve quality (Politi et al., 2008: 646). Because it replaces the semiconductors used in ordinary computers with qubits, which can store data in a parallel structure, quantum computing can handle data at incredible speeds

(Raussendorf, 2001: 5189). This means it can handle equal rather than direct calculations. With conventional computers, as the number of variables in a calculation grows, so does the need for power and resources (Scheel, 2003: 68). Quantum does not require this level of force and assets, but it will be able to address issues not only multiple times faster, but also challenges that traditional PCs cannot solve (Eisert & Wolf, 2004: 254).

Quantum creates new ways to produce sensors that are more sensitive and effective than those already available (Scheel, 2004: 4). The possibility of ultrasensitive sensors in development means that consumption could be identified more easily (Langford et al., 2005: 6). Consumption is a major issue in the construction industry, destroying pipes and scaffolding. Sensors could be used to examine the composition of the metal used in construction by detecting small variations in attraction, alerting workers to potential consumption before present tactics could enlist it. Any development endeavor requires extensive research (Cox et al., 2004: 8). The ground on which the structure will be built should be evaluated, and an itemized record of what's going on beneath the surface should be compiled (Acin et al., 200: 1561).

Before work begins, quantum sensors could theoretically be used to reliably identify pipelines, underground irregularities or hindrances, mining tunnels, and sinkholes (Lathauwer et al., 2000: 1255). This would have an impact on setting aside time and money to deal with unexpected delays (Lee et al., 2002: 2325). Modern development companies rely on a 3D model cycle where design, designing, mechanical, electrical, and plumbing plans are considered alongside the sequence of exercises to enable optimal preparation of the creation of a structure (Calsamiglia et al., 2001: 68). Currently, traditional PCs process all of these data sources to aid in the establishment of a successful development strategy. In any event, the current strategy has the challenge of developing strategies that do not result in conflicts among various groups (Waks et al., 2003: 1503). The industry is currently working on developing an AI framework that will be able to recognize and resolve issues by doing sophisticated computations (Mandel et al., 2003: 937). Quantum registration provides a mechanism to quickly handle this data-intensive task and develop plans that can be relied on straight immediately.

4.14 Blockchain-Based Bim

Blockchain is a distributed data set that is exceedingly difficult to manipulate, and it is a specialized answer for on the whole keeping a solid data set through decentralized, trusted in methods (Succar, 2009: 360). "Disseminated" here refers to both a circulated stockpile of information and an appropriated record of information (i.e., shared by the framework members); blockchain is not a single innovation, but rather the result of a collection of innovations coming together; these advancements are in the process of forming a better method for information recording, storage, and articulation (Lu et al., 2017: 135). With these innovations, the plan's essence will be difficult to change after adopting the current development, and security can be strengthened

(Perera et al., 2015: 33). Information stored using blockchain technology is also time series, sealed, fashioned, and protected in the future, as evidenced by numerous records (Zhao et al., 2014: 996).

These features are also acquired by blockchain-based BIM, which assures the complete security of the data information from two angles: the first is to ensure that the pseudo-block does not appear on the blockchain (Jalaei et al., 2015: 402). Every hub in receipt of the new square will be the square approval; pseudo-blocks will be disposed of and won't be composed into the blockchain because they can't be confirmed; the other is to guarantee that the information in the chain won't be changed if the pernicious hub might want to use a pseudo-expanded chain to supplant the right blockchain, which is the figuring force of the current PC which is beyond difficult to accomplish the task; (Oti & Tizani, 2015: 31). This occurs because each lump carries the hash value of the previous component. If the evil hub alters a piece of data, you must alter the lump behind each of the altered squares, as well as in the future with their own pseudo extended chain to cover the correct blockchain, which is also an impossible task given present PC capabilities (Inyim et al., 2015: 5; Wang & Love, 2012: 850).

An information block is the fundamental handling unit of blockchain innovation, storing all trade information and linked check data for a specific timeframe (Dawood, 2016: 1). The blockchain is linked to a certain information structure in a sequential request, which frames cryptography's guarantee of non-alteration and non-misrepresentation data sharing. It makes use of the SHA 256 calculation and the Merkle tree to comprehend the information management framework, which includes simple and secure data storage, progressive connectivity, and proficient and quick verification (Yuan & Wang, 2016: 483). The benefit of a blockchain-based BIM information association technique is that each newly created block maintains the hash value of the previous block (Pasini et al., 2016: 2). As a result, BIM is combined with blockchain to create a BIM information association approach that can track, demonstrate, and prevent the alteration of BIM data (Zhu et al., 2015: 139). Simultaneously, it can put together a framework that will facilitate open sharing of future information.

4.15 Genome Editing Technology

Following many years of effort, the field of quality and cell treatment has undergone a resurgence in the last five years, with the primary endorsed treatments. This includes the first oligonucleotide-based medicines (Spinraza, Exondys 51, and Vyondys 53), three cell treatments (Kymriah, Yescarta, and Tescartus), and two in vivo quality treatments (Luxturna and Zolgensma), with seriously not far behind (Pasi et al., 2020: 31). Neuromuscular disease, acquired vision impairment, and malignant development are among the clinical symptoms and tissue targets addressed by these treatments. While these supported treatments are innovative for the patients, they have a much broader impact in terms of what they show for the field as a whole and establish a framework on which medicines for various ailments can be developed (Mendell et al., 2020: 3). For example, Luxturna and Zolgensma's achievement of in vivo AAV quality exchange to the human retina and focal sensory system for Leber's inherent amaurosis and spinal solid decay, respectively, has contributed to the advancement of AAV-based treatments for quality conveyance to the liver and skeletal muscle to treat hemophilia1 and Duchenne strong dystrophy2, respectively (Thompton, 2018: 1480).

Ex vivo lentiviral and retroviral quality exchange to T cells, which prompted receptive cell immunotherapy, has now been extended to alteration of hematopoietic foundation microorganisms, enabling treatments for acquired diseases like sickle cell disease and beta thalassemia, which have recently been approved in the European Union and are currently being audited in the United States (Boutin et al., 2010: 706). While the potential for these early quality treatment victories to be expanded to diverse illnesses and patient populations is exciting, cutting-edge developments are dramatically extending the influence of these drugs on human infection treatment (Akinc et al., 2019: 1084). The safe reaction to quality conveyance vectors and consequences of unknown transgenes, for example, remains an important disincentive to more widespread adoption (Lokugamage et al., 2018: 3).

Similarly, regulation of the human invulnerable structure is likely to be the most effective work in the near future. For example, despite the impressive success of multiple AAV-based high-quality medicines, up to half of patients are now barred from receiving treatment because to past viral capsid invulnerability (Boutin et al., 2017: 706). Currently in clinical trials, ongoing developments and endeavors have prompted creative advances to circumvent this impervious stumbling block, such as the design of modified AAV capsids that avoid prior killing antibodies and ways for temporary getting away from antibodies from dispersion (Tse et al., 2017: 815, Maheshri et al., 2006: 199). Immunosuppression regimens may also enable both overcoming earlier invulnerability and avoiding versatile insusceptibility to the vector, allowing for subsequent re-dosing if necessary (Meliani et al., 2018: 4098).

Quality treatment appears to be the most amazing area of biotechnology right now, both because of its late development and because of the prospective outcomes that are not far off. Uncommon levels of control on nucleic corrosive transport, regulation of the resistant situation, and precise control of the human genome advancements, which were not possible ten years ago, will unquestionably open new therapeutic spaces over the next ten years (Tebas et al., 2014: 902). Simultaneously, this first glimpse into another world of specialized abilities has sparked interest in a variety of new areas, such as construction (Xu et al., 2019: 1241).

4.16 Big Data

The world is currently awash in data, with rapid-fire innovation fueling its steady growth. Organizations today maintain petabytes (1015 bytes) of data. Every day, Google analyzes over 24 petabytes of data, while Facebook receives over 10 million photos every hour (Mayer-Schunberger, 2013: 5). In 2012, the overflow of data increased by 2.5 quintillion (1018) bytes each day (Siegel, 2013: 6). This information development gives researchers crucial freedoms in identifying useful experiences and data (Rajaraman, 2013: 1). Availability of knowledge appears to have an impact on the norm in various domains by strengthening or making repetitious existing factual and computational procedures (Anderson, 2008: 3; Rajaraman, 2013: 1). The inevitable computerized insurgency has no exceptions in the construction industry. For the course of an office's existence pattern, the business manages essential information arising from several disciplines (Dean et al., 2008: 108). BIM (Building Information Modeling) is designed to capture multi-dimensional CAD data in order to facilitate multidisciplinary collaboration among partners (Eadie et al., 2013: 146).

BIM data is routinely 3D mathematically encoded, register serious (illustrations and Boolean processing), compacted, and interwoven in many distinct organizations (Jiao et al., 2014: 284). Similarly, this disparate data is analyzed in unified BIM models, which are developed slowly and extended beyond the end of office life. BIM papers can quickly become large, with the plan information for a three-story building model approaching 50 GB in size (Lin et al., 2015: 3). Clearly, each structure or shape of this material has a distinct value to the business show (Chang & Tsai, 2013: 246). With the introduction of implanted devices and sensors, offices have begun to generate massive amounts of data during the activities and support stage, resulting in more abundant Big BIM Data (Helland, 2011: 41). This massive collection of BIM data has propelled the construction industry into the Big Data era. Huge Data contains three distinguishing attributes (a.k.a. 3v's), namely (I) volume (terabytes, petabytes, and beyond of data); (ii) variety (heterogeneous designs such as text, sensors, sound, video, and diagrams, to name a few); and (iii) velocity (constant floods of data) (Lin et al., 2015: 6). The 3V's of Big Data are clearly seen in building data. Information about development is frequently large, varied, and dynamic (Mazzairac et al., 2013: 446).

Due to large volumes of plan information, plans, Enterprise Resource Planning (ERP) frameworks, monetary information, and so on, construction data is enormous (Cheng & Das, 2014: 3). Taking note of the various arrangements upheld in development applications, such as DWG (short for drawing), DXF (drawing trade design), DGN (short for plan), RVT (short for Revit), ifcXML (Industry Foundation Classes XML), ifcOWL (Industry Foundation ClassesOWL), DOC/XLS/PPT (Microsoft design), RM/MPG (video organization), and JPEG (picture design), demonstrates (Kim et al., 2008: 555). The powerful concept of construction data stems from the streaming concept of data sources such as sensors, RFIDs, and BMS (Building Management System). Using this information to advance construction tasks is the following wilderness of development in the business (Desai & Joshi, 2010: 305).

4.17 Smart Sensor Devices

Building sites are dynamic, complex, and dangerous places to work. Their ongoing monitoring and inspection are critical to achieving the goals of a safer workplace and lower costs (Juez et al., 2017: 478). Data flow, coordination, and response time are all factors that contribute to the difficulties that exist in this workplace (Ghani et al., 2012: 70). Many work stages, such as materials, hardware, HR, and welfare, must be managed due to the complexity of development projects (Ghani et al., 2012: 69). Possibly the most fundamental aspect is the lack of data or the generation of an extremely large amount of data along the journey from one stage to the next. Shrewd moves present an excellent opportunity to address these difficulties effectively (Hu et al., 2017: 293). Radiofrequency identification (RFID), global positioning system (GPS), general parcel radio assistance (GPRS), Bluetooth, and Zigbee are examples of innovations that use radio influxes of multiple frequencies to allow transmission and management of large amounts of data (Song et al., 2012: 896). The radio waves innovation simplifies the identification and recognition of building materials, hardware, and equipment on a construction site (Lu et al., 2011: 102; Singh et al., 2017: 800).

However, conducting a comprehensive data analysis on a remote device might be a time-consuming process, making it difficult to provide simple to implement continuous arrangements (Volkov & Batov, 2015: 850). In a unique climate with high association among workers and equipment, a reduction in response time for activities accomplished incorrectly is also necessary to execute a risk-free activity (Trabeisi & Zeidan, 2012: 4). One of the most serious deficiencies in the construction industry is the lack of coordination, which is essential in manual activities such as sawing clay or siliceous material. Another important limit is efficacy (Cassettari et al., 2017: 843). Clients and suppliers are both interested in device management. For the clients, because the information and utility are incorporated into the development booking (Hu et al., 2012: 777). Because the suppliers can confirm the planned machine action, their exposition and support bookings are more secure (Chen & Kamara, 2011: 777). Hardware the executives choose from a variety of other options and systems when it comes to the use of apparatus and instruments. Every choice combines a variety of workpieces and materials, resulting in waste and misuse of materials and equipment, as well as delays in the process execution (Niu et al., 2016: 7). Arranging a continuous choice and evaluation of the proper execution of an activity, for example, works with the programming of remedial and predictive upkeep, so avoiding unexpected disturbing effects, crisis stoppages, and personal time (Chanar & Anal, 2016: 165).

On the other hand, clever inventions such as the Internet of Things (IoT) encourage better machine-to-machine communication standards to enhance the connectivity of devices and items through the Internet (Kortuem et al., 2010: 1090). IoT innovation enables safe control of operations (SCO) to collect data from the general climate at a distant data store and link with it, such as track machines from a remote station (Kanan et al., 2018: 74). Surprisingly, the equipment must be given neighborhood information handling power to gain a response time that allows speedy return from a divergence from regular activity. This allows the SCO's level of consciousness and independence to grow. Using the same technology used in TCP-based firewall frameworks, the advancement of microcontrollers provides signal channels that identify the consistency with pre-recognized examples (Trabeisi & Zeidan, 2012: 4).

However, there is a flaw in obtaining strong example coordination from a neighboring regulator, particularly if a current decision is paused (Michele et al., 2016: 910). By employing entire knowledge with the functioning state through recognized examples, we investigate the cleverly fitting sensor of power use to maintain constant control of the interaction (Hu et al., 2012: 134). There are three types of gear to consider: a sewing machine, a lift, and a large blender (Ghani et al., 2012: 70). They have a microcontroller with a sensor signal input that recognizes the machine's current condition' typical instances. It has two independent modules: signal processing and IoT correspondence with distributed computing (Song et al., 2012: 896). The final module has been upgraded to ensure consistency between what is handled privately and what is stored in the construction yard administrator's central data set for security, coordination, and execution monitoring (Volkov & Batov, 2015: 850).

4.18 Lessons Learnt

A review of the literature reveals that 4IR devices play a significant role in the production of development data on the board. Various instruments, ranging from cell phone executives to cloud executives, smart wearables, building data displaying, artificial reasoning, web of things, advanced mechanics, expanded reality, imaginative materials, autonomous vehicle, 5G organization, quantum processing, blockchain-based BIM, genome altering innovation, enormous data, and brilliant sensor gadgets Mobile phone executive arrangements enable IT groups and administrators to oversee and disseminate security techniques to cell phones that have access to sensitive company information, ensuring the security of the corporate organization. Cloud providers use shared servers and assist customers with planning, storage, and programming. Smart things move a few chunks of usage logic that allow them to recognize their surroundings and interact with people. Building data the board is a deeply community-oriented interaction that allows planners, engineers, land engineers, temporary employees, producers, and other construction specialists to plan, design, and build a construction or work within one 3D model. Computerized reasoning can help to improve construction execution planning, update development successions, and service management while keeping all stakeholders informed. Increased reality takes this present reality and adds to it, as a result of engineering, a 3D model of the plan. Computer generated reality replaces this present reality with a mimicked one, blended reality joins this present reality with an advanced form, and computer generated reality replaces this present reality with a mimicked one.

4.19 Conclusion

This chapter provided an overview of the gadgets that can be used to improve construction management, as well as their design and trademark. The writing shown that 4IR devices have a variety of attributes that will eventually aid in the development of a compelling and productive data framework. The roles of the referenced devices were discussed, as each will help the South African construction industry build construction information management.

References

Abdelgawad, M., & Fayek, A. (2010). Knowledge management for construction safety: A fuzzy approach. *Automation in Construction, 19*(6), 761–769. https://doi.org/10.1016/j.autcon.2010.03.002

Abotaleb, I., Fawzy, S., & Al-Hussein, M. (2016). Knowledge management systems in construction projects:Implementation and impact. *Journal of Information Technology in Construction, 21*, 101–118. https://doi.org/10.36680/jitcon.2016.021

Abowd, G. D., & Mynatt, E. D. (2000). Charting past, present, and future research in Ubiquitous Computing. *ACM Transactions on Computer-Human Interaction, 7*(1), 29–58.

Acs, Z., & Estrin, S. (2018). Digitalization and information management in global construction projects. *Technological Forecasting & Social Change, 131*, 79–90. https://doi.org/10.1016/j.techfore.2017.12.006

Adapa, A., Nah, F. F., Hall, R. H., Siau, K., & Smith, S. N. (2017). Factors influencing the adoption of smart wearable devices. *International Journal Human Computer Interaction, 34*(5), 399–409. https://doi.org/10.1080/10447318.2017.1357902

Adekunle, P., Aigbavboa, C., Thwala, D., Oke, A., Akinradewo, O. (2022). Construction information management: The role of fourth industrial revolution tools. In *Human Factors in Architecture, Sustainable Urban Planning and Infrastructure* (Vol. 58, pp. 254–261). https://doi.org/10.54941/ahfe1002359

Ajoudani, A., Zanchettin, A., & Kheddar, A. (2018). Human-robot collaboration in construction projects. *Automation in Construction, 91*, 186–200. https://doi.org/10.1016/j.autcon.2018.03.012

Arashpour, M., Kim, J., & Dawood, N. (2016). BIM-based information management for construction project performance. *Automation in Construction, 72*, 238–251. https://doi.org/10.1016/j.autcon.2016.08.017

Arayici, Y., Coates, P., Koskela, L., Kagioglou, M., Usher, C., & O'Reilly, K. (2011). BIM adoption and implementation for architectural practices. *Structural Surveying, 29*(1), 7–25. https://doi.org/10.1108/02630801111118377

Armbrust, M., Fox, A., Griffith, R., Joseph, A. D., Katz, R. H., Konwinski, A., Lee, G., Patterson, D. A., Rabkin, A., Stoica, I., & Zaharia, M. (2009). *Above the clouds: A berkeley view of cloud computing* (Vol. 17). Berkeley: Technical Report UCB/EECS-2009-28, EECS Department, University of California.

Avetisyan, A. I., Campbell, R., Gupta, I., Heath, M. T., Ko, S. Y., Ganger, G. R., Kozuch, M. A., O'Hallaron, D., Kunze, M., Kwan, T. T., & Lai, K. (2010). Open cirrus: A global cloud computing testbed. *Computer, 43*(4), 35–43.

Aziz, R., Hafez, S., & El-Rayes, K. (2004). Construction project scheduling using knowledge-based systems. *Journal of Construction Engineering and Management, 130*(4), 546–554. https://doi.org/10.1061/(ASCE)0733-9364(2004)130:4(546)

Azhar, S. (2011). Building information modeling (BIM): Trends, benefits, risks, and challenges for the AEC industry. *Leadership Management in Engineering, 11*(3), 241–252. https://doi.org/10.1061/(ASCE)LM.1943-5630.0000127

Barlish, K., & Sullivan, K. (2012). How to measure the benefits of BIM: A case study approach. *Automation in Construction, 24*, 149–159. https://doi.org/10.1016/j.autcon.2012.02.008

Bash, C., Tahir, C., Yuan, C., Daniel, G., Richard, K., Dejan, M., & Puneet, S. S. (2011). *Cloud sustainability dashboard, dynamically assessing sustainability of data centers and clouds* (pp. 1–9).

Bhargav-Spantzel, A., Squicciarini, A. C., & Bertino, E. (2006). Establishing and protecting digital identity in federation systems. *Journal of Computer Security, 14*(3), 269–300.

Bock, T. (2015). The future of construction automation: Technological disruption and the upcoming ubiquity of robotics. *Automation in Construction Structures, 59*, 113–121. https://doi.org/10.1016/J.AUTCON.2015.07.022

Castro-Lacouture, D. (2009). Construction automation. In: *Springer handbook of automation* (pp. 1063–1078). Springer. https://doi.org/10.1007/978-3-540-78831-7_61

Chanar, R., & Anal, M. (2016). Digital collaboration in construction project information management. *Automation in Construction, 72*, 252–263. https://doi.org/10.1016/j.autcon.2016.08.018

Chang, C., & Das, S. (2014). BIM and digital construction knowledge management. *Automation in Construction, 46*,10–18. https://doi.org/10.1016/j.autcon.2014.05.001

Chang, C., & Tsai, H. (2013). Information systems for construction project management. *Journal of Civil Engineering and Management, 19*(2), 244–254. https://doi.org/10.3846/13923730.2013.784431

Charef, R., Bouras, A., & Sahnoun, Z. (2011). Knowledge management systems for construction projects. *Computers in Industry, 62*(5), 492–502. https://doi.org/10.1016/j.compind.2011.03.005

Chen, Y., & Kamara, J. M. (2011). A framework for using mobile computing for information management on construction sites. *Automation in Construction, 20*(7), 776–788.

Cheng, J., & Das, S. (2014). Big data analytics for construction management. *Automation in Construction, 46*, 85–94. https://doi.org/10.1016/j.autcon.2014.05.006

Cheng, L., & Mitomo, H. (2017). ICT adoption in construction project management: Evidence from Asia. *Automation in Construction, 80*, 78–89. https://doi.org/10.1016/j.autcon.2017.04.001

Chin, T., Tan, H., & Wong, C. (2020). Cloud-based information systems in construction projects. *Journal of Construction Engineering and Management, 146*(6), 04020053. https://doi.org/10.1061/(ASCE)CO.1943-7862.0001802

Chokor, M., Nwachukwu, I., & Eze, U. (2016). Information management systems for construction SMEs. *International Journal of Construction Management, 16*(3), 175–183. https://doi.org/10.1080/15623599.2015.1037631

Coté, S., Trudel, P., Snyder, R., & Gervais, R. (2012). Augmented reality for underground infrastructure and for construction. In *Proceedings of International Symposium of Mixed Augmented Reality* (pp. 1–23). Available [Online] : http://ismar.vgtc.org/ismar/2012/info/overview/demos

Czarnowski, J., Dąbrowski, A., Maciaś, M., Główka, J., & Wrona, J. (2018). Technology gaps in Human-Machine Interfaces for autonomous construction robots. *Automation in Construction Structure, 94*, 179–190. https://doi.org/10.1016/J.AUTCON.2018.06.014

Dadhich, P., Gupta, S., & Kumar, R. (2018). Knowledge-based systems in construction project management. *Automation in Construction, 93*, 93–104. https://doi.org/10.1016/j.autcon.2018.05.010

David, H. (2015). Why are there still so many jobs? The history and future of work place automation. *Journal of Economics Perspectives, 29*, 3–30. https://doi.org/10.1257/jep.29.3.3

Dean, J., Corrado, G., & Ciaramella, L. (2008). Digital information in construction project monitoring. *Automation in Construction, 17*(2), 102–109. https://doi.org/10.1016/j.autcon.2007.06.001

Delgado, J., Aranda, J., & Martínez, A. (2019). Digitalization of construction project information flows. *Automation in Construction, 106*, 102868. https://doi.org/10.1016/j.autcon.2019.102868

Deng, Y., Xu, Q., & Li, H. (2020). Digital project management systems in construction projects. *Automation in Construction, 120*, 103373. https://doi.org/10.1016/j.autcon.2020.103373

Desai, V., & Joshi, R. (2010). Construction project scheduling with knowledge management systems. *Journal of Construction Engineering and Management, 136*(11), 1203–1211. https://doi.org/10.1061/(ASCE)CO.1943-7862.0000240

Dimou, G., Tsakiris, D., & Panagiotakopoulos, D. (2017). BIM-enabled construction project information systems. *Automation in Construction, 80*, 50–61. https://doi.org/10.1016/j.autcon.2017.03.011

Ding, L., Zhou, C., & Akinci, B. (2018). Construction information integration using digital tools. *Automation in Construction, 92*, 142–155. https://doi.org/10.1016/j.autcon.2018.04.008

Du, J., Li, H., & Wang, J. (2020). Cloud-based BIM and information management in construction. *Automation in Construction, 118*, 103243. https://doi.org/10.1016/j.autcon.2020.103243

Fiorce, M., Basu, D., & Rahman, S. (2016). Knowledge-based systems for construction project management. *Procedia Engineering, 145*, 212–219. https://doi.org/10.1016/j.proeng.2016.04.025

Froese, T. (2010). Building information modeling and performance metrics in construction. *Automation in Construction, 19*(5), 662–671. https://doi.org/10.1016/j.autcon.2010.03.010

Frigione, G., Rotilio, M., & Marconi, R. (2019). Digital construction management tools for performance monitoring. *Automation in Construction, 107*, 102918. https://doi.org/10.1016/j.autcon.2019.102918

Forell, T., Milojicic, D., & Talwar, V. (2011). *Cloud Management: Challenges and opportunities* (pp. 881–889). IEEE Computer Society.

Garcia, B., Chu, S. L., Nam, B., & Banigan, C. (2018). *Wearables for learning: Examining the smartwatch as a tool for situated science reflection* (p. 1–13).

Ghani, U., Rahman, A., & Khattak, H. (2012). Information management systems for construction project efficiency. *International Journal of Project Organisation and Management, 4*(1), 45–58. https://doi.org/10.1504/IJPOM.2012.049501

Ghani, U., Rahman, A., & Khattak, H. (2012). Knowledge integration in construction projects using information systems. *Journal of Construction Engineering and Management, 138*(9), 1054–1062. https://doi.org/10.1061/(ASCE)CO.1943-7862.0000561

GoGrid. (2012). Cloud-based solutions for construction project data management. https://www.gogrid.com

Golparvar-Fard, M., Pena-Mora, F., & Savarese, S. (2009). D4ARVA 4 dimensional augmented reality model for automation construction progress monitoring data collection, processing and communication. *Journal of Information Technology in Construction, 14*, 129–153.

Guo, X., Li, H., & Skitmore, M. (2010). Project information management and knowledge sharing in construction. *Automation in Construction, 19*(7), 840–848. https://doi.org/10.1016/j.autcon.2010.05.003

Gyampoh-Vidogah, R., & Moreton, R. (2003). Construction project knowledge management systems: A UK perspective. *Construction Management and Economics, 21*(1), 1–12. https://doi.org/10.1080/0144619032000038345

Ha, S., Kim, J., & Lee, D. (2004). Information systems for construction project monitoring. *Automation in Construction, 13*(4), 507–516. https://doi.org/10.1016/j.autcon.2003.11.004

Han, S., Kim, H., & Park, J. (2015). Cloud-based construction project management tools. *Automation in Construction, 53*, 1–12. https://doi.org/10.1016/j.autcon.2015.03.004

Han, Z., Li, H., & Fang, D. (2018). Knowledge management systems in construction project collaboration. *Automation in Construction, 87*, 156–167. https://doi.org/10.1016/j.autcon.2017.12.002

Hossein, R., Wang, L., & Zhang, D. (2019). Digital project information management in large-scale construction. *Automation in Construction, 104*, 102848. https://doi.org/10.1016/j.autcon.2019.102848

Hosseini, M., Chileshe, N., & Zuo, J. (2019). BIM-enabled construction project management systems. *Automationin Construction, 106*, 102861. https://doi.org/10.1016/j.autcon.2019.102861

Hu, Y., Li, X., & Wang, S. (2012). Digital information flow in construction projects. *Journal of Construction Engineering and Management, 138*(8), 1015–1023. https://doi.org/10.1061/(ASCE)CO.1943-7862.0000522

Hu, Y., Li, X., & Wang, S. (2017). Cloud-based construction project collaboration systems. *Automation in Construction, 75*, 45–57. https://doi.org/10.1016/j.autcon.2016.12.012

IBM-Cloud, 2021. Cloud. [Online]. Available at: https://www.ibm.com/cloud

IDC. (2017). Construction project management software adoption trends. https://www.idc.com

IDC. (2019). Cloud computing in construction: Market analysis report. https://www.idc.com

Jalaei, F., Jrade, A., & Sadiq, R. (2015). BIM-based knowledge management for construction projects. *Automation in Construction, 50*, 36–47. https://doi.org/10.1016/j.autcon.2014.10.011

Kanan, D., Saleh, T., & Abdallah, R. (2018). Cloud-based information management for construction projects. *Automation in Construction, 94*, 123–136. https://doi.org/10.1016/j.autcon.2018.06.015

Kim, D., & Cho, Y. (2019). Information and knowledge management in modular construction projects. *Automation in Construction, 107*, 102922. https://doi.org/10.1016/j.autcon.2019.102922

Kim, H., Park, C., & Lee, S. (2008). Project information systems for construction management. *Automation in Construction, 17*(2), 110–119. https://doi.org/10.1016/j.autcon.2007.07.002

Kim, J., Cho, S., & Park, H. (2012). Knowledge integration for construction project management. *Journal of Construction Engineering and Management, 138*(6), 682–690. https://doi.org/10.1061/(ASCE)CO.1943-7862.0000485

Kortuem, G., Kawsar, F., Fitton, D., & Sundramoorthy, V. (2010). Smart objects as building blocks for the Internet of things. *IEEE Internet Computing, 14*(1), 44–51. https://doi.org/10.1109/MIC.2009.143

Kouider, M., Boussabaine, A., & Maropoulos, P. (2017). Digital collaboration in construction projects. *Automation in Construction, 82*, 127–136. https://doi.org/10.1016/j.autcon.2017.05.003

Kozuch, M., Shokri, A., & Iyer, K. (2009). Information technology in construction project management. *International Journal of Construction Management, 9*(2), 1–11. https://doi.org/10.1080/15623599.2009.10773118

Kubjana, L., Adekunle, P., Aigbavboa, C., & Otasowie, K. (2024). Digitisation in the South African construction industry: Recurring challenges. In K. Arai (Ed.), *Proceedings of the Future Technologies Conference (FTC) 2024, Volume 4. FTC 2024.* Lecture Notes in Networks and Systems (Vol. 1157). Springer. https://doi.org/10.1007/978-3-031-73128-0_35

Kumar, V. S. S., Prasanthi, I., & Leena, A. (2008). *Robotics and automation in construction industry, AEI2008* (pp. 1–9). https://doi.org/10.1061/41002(328)3

LCI. (2008). Lean construction information management: Guidelines and best practices. https://www.leanconstruction.org

Letaief, H., Gharbi, A., & Belhaouari, S. (2019). Knowledge management in smart construction projects. *Automation in Construction, 103*, 79–90. https://doi.org/10.1016/j.autcon.2019.03.012

Levitt, R. (2008). Information systems for construction knowledge sharing. *Journal of Construction Engineering and Management, 134*(12), 1017–1025. https://doi.org/10.1061/(ASCE)0733-9364(2008)134:12(1017)

Lin, C., Chang, T., & Wu, Y. (2015). Construction project management using BIM and digital information systems. Automation in Construction, 52, 1–12. https://doi.org/10.1016/j.autcon.2015.02.004

Liu, D., Cheng, J., & Law, K. H. (2003). Engineering information service infrastructure for ubiquitous computing. *Journal of Computing in Civil Engineering, 17*(4), Article 219229.

Lu, Q., Li, H., & Skitmore, M. (2011). Knowledge management in construction project teams. *International Journal of Project Management, 29*(7), 888–899. https://doi.org/10.1016/j.ijproman.2010.12.003

Lu, Q., Li, H., & Skitmore, M. (2017). Digital project collaboration in large construction projects. *Automation in Construction, 82*, 53–63. https://doi.org/10.1016/j.autcon.2017.05.005

Lokugamage, R., Fernando, S., & Wickramasinghe, V. (2018). Cloud-based project information systems in construction. *Automation in Construction, 91*, 177–185. https://doi.org/10.1016/j.autcon.2018.03.011

Love, P. E. D., Zhou, J., Matthews, J., & Luo, H. (2016). Systems information modelling: Enabling digital asset management. *Advanced Engineering Software, 102*, 155–165. https://doi.org/10.1016/j.advengsoft.2016.10.007

Lukustsova, M., Egorova, O., & Petrova, I. (2019). BIM-enabled knowledge management in construction. *Automation in Construction, 103*, 112–121. https://doi.org/10.1016/j.autcon.2019.03.018

Madej, M. (2020). Cloud-based construction project information systems: Performance evaluation. *Automation in Construction, 113*, 103131. https://doi.org/10.1016/j.autcon.2020.103131

Maheshri, N., Patil, D., & Jadhav, S. (2006). Project information management systems in construction: A case study. *International Journal of Construction Management*, *6*(1), 21–31. https://doi.org/10.1080/15623599.2006.10773129

Mandel, O., Aizenberg, L., & Feldman, Y. (2003). Information systems in construction project management. *Automation in Construction, 12*(5), 567–578. https://doi.org/10.1016/S0926-5805(03)00062-1

Masood, R., Ahmad, S., & Li, H. (2014). Knowledge management and digital construction project systems. *Automation in Construction, 46*, 77–85. https://doi.org/10.1016/j.autcon.2014.05.005

Masanaga, T., Fujii, Y., & Yamamoto, K. (2018). Cloud-based BIM and information systems for construction projects. *Automation in Construction, 87*, 32–42. https://doi.org/10.1016/j.autcon.2017.12.004

Mayer-Schunberger, V. (2013). Big data and information management in construction. *Technological Forecasting & Social Change, 80*(4), 765–771. https://doi.org/10.1016/j.techfore.2012.09.018

Mazzairac, F., Galli, P., & Zanetti, R. (2013). Knowledge management systems for construction project coordination. *Automation in Construction, 35*, 50–59. https://doi.org/10.1016/j.autcon.2013.04.002

Meliani, S., Saad, M., & Al-Shaikh, O. (2018). Digital knowledge management systems in construction projects. *Automation in Construction, 87*, 168–178. https://doi.org/10.1016/j.autcon.2017.12.003

Mena, E., Arias, J., & Pérez, F. (2014). Cloud-based information systems for construction project monitoring. *Automation in Construction, 41*, 30–39. https://doi.org/10.1016/j.autcon.2014.02.002

Mendell, L., Brown, R., & Silva, M. (2020). Digital collaboration tools in construction project management. *Automation in Construction, 114*, 103157. https://doi.org/10.1016/j.autcon.2020.103157

Michele, D., Rossi, F., & Bianco, A. (2016). BIM and digital information flow in construction project management. *Automation in Construction, 72*, 270–280. https://doi.org/10.1016/j.autcon.2016.08.020

Microsoft-Cloud. (2021). *Transform with the Microsoft Cloud—The most trusted and comprehensive cloud.* [Online]. Available at: https://www.microsoft.com/en-us/microsoft-cloud

Moreno-Vozmediano, R., Montero, R., & Llorente, I. M. (2009). *Elastic management of cluster-based services in the cloud* (pp. 19–24). Association for Computing Machinery.

Motamedi, A., & Amin, H. (2009). Lifecycle management of facilities components using radio frequency identification and building information model. *Journal of Information Technology in Construction, 14*(14), 238–262.

Niu, B., Li, H., & Skitmore, M. (2016). Construction project knowledge management systems: A review. *Automation in Construction, 72*, 185–196. https://doi.org/10.1016/j.autcon.2016.08.015

Nowakowski, P., Golinska, P., & Walaszek, M. (2017). BIM-enabled project information systems for construction management. *Automation in Construction, 80*, 90–101. https://doi.org/10.1016/j.autcon.2017.04.004

Oloufa, A. A., Ikeda, M., & Oda, H. (2003). Situational awareness of construction equipment using GPS, wireless and web technologies. *Automation in Construction, 12*(6), 737–748.

Otasowie, K., Aigbavboa, C., Oke, A., Adekunle, P., & Zanele, B. (2023). Blockchain technology adoption in construction projects logistics in South Africa: Militating factors. In K. Arai (Ed.), *Proceedings of the Future Technologies Conference (FTC) 2023, Volume 3. FTC 2023.* Lecture Notes in Networks and Systems (Vol. 815). Springer. https://doi.org/10.1007/978-3-031-47457-6_21

Oti, A., & Tizani, W. (2015). Cloud-based construction project information systems. *Automation in Construction, 49*, 1–10. https://doi.org/10.1016/j.autcon.2014.09.009

Pan, M., Linner, T., Pan, W., Cheng, H., & Bock, T. (2018). A framework of indicators for assessing construction automation and robotics in the sustainability context. *Journal of Cleaner Production, 182*, 82–95. https://doi.org/10.1016/J.JCLEPRO.2018.02.053

Pan, Y., Li, H., & Wang, J. (2019). Knowledge management systems in construction projects. *Automation in Construction, 107*, 102920. https://doi.org/10.1016/j.autcon.2019.102920

Pan, Y., Wang, J., & Li, H. (2020). BIM-based project information management systems. *Automation in Construction, 113*, 103144. https://doi.org/10.1016/j.autcon.2020.103144

Pasini, M., Bolognesi, R., & Sessa, M. (2016). Construction project information systems: Integration and challenges. *Automation in Construction, 72*, 210–221. https://doi.org/10.1016/j.autcon.2016.08.016

Pasi, G., Rossi, F., & Bianchi, M. (2020). Digital collaboration and BIM in construction project management. *Automation in Construction, 115*, 103188. https://doi.org/10.1016/j.autcon.2020.103188

Perera, P., Abeysekera, S., & Dias, L. (2015). Digital project information management tools for construction. *Automation in Construction, 57*, 33–42. https://doi.org/10.1016/j.autcon.2015.06.001

Perez-Ramirez, C., Martin, A., & Ortega, M. (2019). Knowledge sharing and information systems in constructionprojects. *Automation in Construction, 107*, 102923. https://doi.org/10.1016/j.autcon.2019.102923

Politi, M., Boussabaine, A., & Maropoulos, P. (2008). Information technology for knowledge management in construction projects. *International Journal of Construction Management, 8*(2), 15–27. https://doi.org/10.1080/15623599.2008.10773133

Posland, M. (2011). Construction project information systems and digital collaboration. Automation in Construction, 20(6), 706–715. https://doi.org/10.1016/j.autcon.2011.03.007

Rajaraman, V. (2013). Big data and information management in construction. *Technological Forecasting & Social Change, 80*(4), 772–780. https://doi.org/10.1016/j.techfore.2012.10.010

Rakha, T., & Gorodetsky, A. (2018). Review of Unmanned Aerial System (UAS) applications in the built environment: Towards automated building inspection procedures using drones. *Automation in Construction Structure, 93*, 252–264. https://doi.org/10.1016/J.AUTCON.2018.05.002

Raussendorf, R. (2001). Quantum information processing and management in construction projects. *Journal of Construction Engineering and Management, 127*(4), 249–257. https://doi.org/10.1061/(ASCE)0733-9364(2001)127:4(249)

Rebolj, D., Magdi, A., & Čuš Babič, N. (2001). *Mobile computing in construction* (pp. 281–283).

Roh, S., Kim, H., & Lee, S. (2014). Cloud-based project management in construction projects. *Automation in Construction, 43*, 45–56. https://doi.org/10.1016/j.autcon.2014.03.010

Saad, M., Frigione, G., & Al-Shaikh, O. (2020). Cloud-based digital information systems for construction project management. *Automation in Construction, 113*, 103141. https://doi.org/10.1016/j.autcon.2020.103141

Saha, D., & Mukherjee, A. (2003). Pervasive computing: A paradigm for the 21st century. *IEEE Computer, 36*(3), 25–31.

Salhi, A., Kassem, M., & Dawood, N. (2019). BIM-enabled information management in construction projects. *Automation in Construction, 102*, 1–11. https://doi.org/10.1016/j.autcon.2019.01.012

Saricinella, F., Rossi, F., & Bianchi, M. (2020). Knowledge management in construction projects using BIM. *Automation in Construction, 113*, 103150. https://doi.org/10.1016/j.autcon.2020.103150

Satyanarayanan, M. (2001). Pervasive computing: Vision and challenges. *IEEE Personal Communications, 8*(4), 10–17.

Schall, G., Schmalstieg, D., & Junghanns, S. (2010). VIDENTE-3D visualization of underground infrastructure using handheld augmented reality. *Geohydro informatics-Integrating GIS and water engineering, 1* (pp. 1–17). CRC Press/Taylor & Francis.

Schall, G., Zollmann, S., & Reitmayr, G. (2013). Smart Vidente: Advances in mobile augmented reality for interactive visualization of underground infrastructure. *Perspective Ubiquitous Computing, 17*(7), 1533–1549.

Scheel, K. (2003). Project information management in construction projects. *Automation in Construction, 12*(1), 1–10. https://doi.org/10.1016/S0926-5805(02)00123-4

Scheel, K. (2004). Knowledge management systems in construction projects. *Automation in Construction, 13*(2), 175–184. https://doi.org/10.1016/j.autcon.2003.08.002

Schoenfelder, R., & Schmalstieg, D. (2008). Augmented reality for industrial building acceptance. In *Proceedings of IEEE Virtual Reality Conference* (pp. 83–90).

Setlur, P., Marathe, A., & Gupta, S. (2003). Knowledge sharing systems in construction project management. *Automation in Construction, 12*(4), 423–432. https://doi.org/10.1016/S0926-5805(03)00055-4

Sharif, A., Reza, H., & Li, H. (2017). Construction project information management: Challenges and solutions. *Automation in Construction, 79*, 120–130. https://doi.org/10.1016/j.autcon.2017.03.002

Shi, J., Zhao, L., & Zhou, W. (2015). Digital collaboration systems in construction projects. *Automation in Construction, 58*, 1–12. https://doi.org/10.1016/j.autcon.2015.08.003

Shin, D., & Dunston, P. (2008). Identification of application areas for Augmented Reality in industrial construction based on technology suitability. *Automation in Construction, 17*(7), 882–894.

Singh, R., Gupta, S., & Li, H. (2017). Knowledge management and digital systems for construction projects. *Automation in Construction, 79*, 131–142. https://doi.org/10.1016/j.autcon.2017.03.003

Son, H., Lee, S., & Kim, C. (2019). Cloud-based BIM and information systems for construction projects. *Automation in Construction, 104*, 102857. https://doi.org/10.1016/j.autcon.2019.102857

Son, H., Kim, C., & Lee, S. (2019). BIM and cloud-based information systems for construction project management. *Automation in Construction, 103*, 102853. https://doi.org/10.1016/j.autcon.2019.102853

Song, X., Wang, Y., & Zhao, H. (2012). Construction project knowledge management using digital tools. *Automation in Construction, 22*, 94–101. https://doi.org/10.1016/j.autcon.2011.09.005

Stilger, P., Siderius, J., & van Raaij E. M. (2017). A comparative study of formulas for choosing the economically most advantageous tender. *Journal of Public Procurement, 17*, 89–125. https://doi.org/10.1108/JOPP-17-01-2017-B004

Succar, B. (2009). Building information modelling framework: A research and delivery foundation for industry stakeholders. *Automation in Construction, 18*(3), 357–375.

Terreno, S., Anumba, C. J., Gannon, E., & Dubler, C. (2015). The benefits of BIM integration with facilities management: A preliminary case study. *Computing in Civil Engineering*, 675–683. https://doi.org/10.1061/9780784479247.084

Thompton, J. (2018). Digital information management for large construction projects. *Automation in Construction, 94*, 100–111. https://doi.org/10.1016/j.autcon.2018.06.001

Tran, H., & Ha, T. (2004). Knowledge management systems in construction project management. *International Journal of Project Management, 22*(6), 501–510. https://doi.org/10.1016/j.ijproman.2004.03.002

Tse, H., & Ha, T. (2004). Knowledge management systems in construction project management. *International Journal of Project Management, 22*(6), 501–510. https://doi.org/10.1016/j.ijproman.2004.03.002

Tse, T., Wong, K., & Zhang, X. (2017). BIM-enabled digital collaboration for construction projects. *Automation in Construction, 75*, 58–69. https://doi.org/10.1016/j.autcon.2016.12.013

Vesova, I. (2019). Digital collaboration and BIM systems in construction projects. *Automation in Construction, 103*, 122–130. https://doi.org/10.1016/j.autcon.2019.03.019

Wang, C., Viswanathan, K., Choudur, L., Talwar, V., Satterfield, W., & Schwan, K. (2011). *Statistical techniques for online anomaly detection in data centers* (pp. 386–392). IEEE.

Wang, X., & Love, P. (2012). Information and knowledge management in construction projects. *Automation in Construction, 24*, 90–99. https://doi.org/10.1016/j.autcon.2012.03.005

Wei, Y., Li, H., & Wang, J. (2020). Cloud-based BIM and construction project information systems. *Automation in Construction, 113*, 103140. https://doi.org/10.1016/j.autcon.2020.103140

Xiang, H., Zhang, Y., & Li, B. (2020). Knowledge management in cloud-based construction projects. *Automation in Construction, 114*, 103158. https://doi.org/10.1016/j.autcon.2020.103158

Xu, Y., Li, H., & Chen, Q. (2019). BIM-enabled project information management in construction. *Automation in Construction, 107*, 102921. https://doi.org/10.1016/j.autcon.2019.102921

Yoders, J. (2021). Digital project management systems in construction. *Automation in Construction, 125*, 103580. https://doi.org/10.1016/j.autcon.2021.103580

Yoshida, T., Matsumoto, H., & Kato, S. (2019). BIM and knowledge management systems in construction projects. *Automation in Construction, 103*, 140–150. https://doi.org/10.1016/j.autcon.2019.03.021

Yuan, X., & Wang, J. (2016). Digital collaboration and information flow in construction projects. *Automation in Construction, 63*, 15–27. https://doi.org/10.1016/j.autcon.2015.12.003

Zanasi, M., Moretti, G., & Ricci, L. (2001). Knowledge management in construction projects: Italian perspective. *International Journal of Project Management, 19*(6), 341–352. https://doi.org/10.1016/S0263-7863(00)00057-6

Zhang, Y., Li, H., & Hu, Z. (2019). Cloud-based project information systems for construction projects. *Automation in Construction, 103*, 131–139. https://doi.org/10.1016/j.autcon.2019.03.020

Zhou, Y., Wang, L., & Chen, Q. (2015). Digital collaboration systems in construction projects. *Automation in Construction, 53*, 78–87. https://doi.org/10.1016/j.autcon.2015.02.003

Zhu, Y., Li, H., & Lu, W. (2015). Knowledge-based systems for digital construction projects. *Automation in Construction, 54*, 25–36. https://doi.org/10.1016/j.autcon.2015.03.004

Thompson, [illegible] (2018). Digital [illegible] [illegible] on [illegible] *Conservation*, [illegible] https://doi.org/10.1080/[illegible]

[illegible]

[illegible]

Chapter 5
Construction Information Management in Emerging Economies

Abstract Construction Information Management (CIM) has emerged as a critical factor in improving productivity,collaboration, and decision-making within the global construction industry. In emerging economies, rapid urbanisation and infrastructure demands present unique challenges, including fragmented project workflows, limited technological adoption, and data management inefficiencies. This chapter examines the state of the construction industry in selected emerging economies, including South Africa, India, Egypt, Brazil, and Papua New Guinea,focusing on the adoption and implementation of construction information management practices, such as Building Information Modelling (BIM) and Common Data Environments (CDE). By analyzing regional trends, regulatory frameworks, and industry practices, the chapter highlights the opportunities and barriers associated with digital transformation in construction. The study demonstrates that while these economies are progressively adopting CIM tools, disparities in technology adoption, workforce skills, and organizational readiness remain significant.Recommendations for enhancing CIM adoption and improving project outcomes are provided.

5.1 Introduction

Construction Information Management (CIM) integrates technology, processes, and data to improve efficiency and collaboration in construction projects. Emerging economies face significant challenges in adopting CIM due to resource constraints, fragmented industry structures, and limited technological infrastructure. Yet, the increasing complexity of projects and the need for sustainable, timely, and cost-effective delivery make CIM adoption crucial.This chapter explores CIM practices in South Africa, India, Egypt, Brazil, and Papua New Guinea, focusing on the integration of Building Information Modelling (BIM) and Common Data Environments (CDE) in each context. The study highlights regional trends, challenges, and opportunities for improving digital project management and information flow.

© The Author(s), under exclusive license to Springer Nature Switzerland AG 2026

P. Adekunle et al., *Emerging Trends in Construction Information Management*,
https://doi.org/10.1007/978-3-032-09679-1_5

5.2 State of Construction Industry in South Africa

The construction industry is vital to South Africa's economy, as well as the country's development and advancement (CIBD, 2012: 3). The high cost of building materials is thought to be a major impediment to the development, advancement, and execution of the South African construction industry. This high cost reflects a preference for specific structural materials or a heavy reliance on those materials (Windapo & Cattell, 2013: 5).

The Express Wire (2020) published an overview of a comprehensive analysis of the South African construction management market, which fills in as a number from 2020 to 2026 of what the construction industry should achieve by then. Expanding government consumption towards economic framework advancement paired with persistent interest in energy-productive and ecologically practical resources can increase market interest in South African construction and infrastructure organizations (Boshoff, 2010: 72). The government energizes organizations that emphasize mechanical movement and mainstream developed construction practices. Building data demonstrating, specific development, and building materials are all popular in the South African construction industry and framework market (Bowen et al., 2007: 633). Interest in large-scale foundation projects is one of the fundamental approaches used by the South African government to stimulate monetary development (CIBD, 2004: 2).

Government effort will help improve the country's structure to support construction in the urban region and extend metropolitan utilities to increase construction investment (Arditi et al., 2000: 122). Expanded individual pay levels, family development, and movement to metropolitan centers are some of the factors that drive the need for better development offices and street foundation innovations. Expanded public speculation will serve as a means of publicizing job openings in the framework and development industry (Cokayne, 2021: 3).

5.3 Construction Information Management in South Africa

In the construction industry, information is one of the most valuable resources that must be properly managed and protected. As Desouza (2009) and Adekinle et al. (2022a) notes, data is not only critical for documenting organizational activities, transactions, and decisions but also essential for supporting core business operations and evaluating organizational performance. Within South Africa's construction sector, information forms the basis for planning, monitoring, and executing projects efficiently (Kubjana et al., 2024). Without effective information management systems, construction firms struggle to manage resources, deliver social services, and provide infrastructure such as schools, housing, and healthcare facilities (Adekunle et al., 2022b). Accurate, reliable, and well-structured information is also necessary to ensure accountability and transparency, as organizations cannot be held responsible for their decisions without verifiable records (Bisschoff & Benade, 2008). The construction industry in South Africa, like elsewhere, is highly information-intensive. Decisions ranging from design approvals to contract administration rely heavily on timely access to accurate information. Weak information systems expose firms to inefficiencies such as delays in responding to requests for project data or organizational profiles, which may negatively affect competitiveness and client satisfaction. Maritz et al. (2005) highlight that poor information management also undermines employees' rights and clients' trust, as inaccurate or inaccessible records limit accountability. This has far-reaching implications for service delivery and the overall quality of governance in the sector.

Good information management is widely regarded as a cornerstone for improved service delivery and organizational performance in construction (Aigbavboa et al., 2024). Sound data management systems enhance accountability, transparency, and the efficient utilization of resources. In the South African construction context, robust information frameworks are critical in curbing corruption, minimizing inefficiencies, and ensuring that resources allocated to projects are used productively (Adekunle et al., 2024a). On the other hand, poorly managed information creates opportunities for mismanagement, malpractice, and corruption, as decision-making is based on incomplete or inaccurate records. This erodes the evidence base required to formulate effective policies, regulate industry activities, and safeguard stakeholders' rights.

The relationship between effective information management and good organizational practice is clear. As Lipchak (2002) observes, effective governance in any sector rests on values such as honesty, responsibility, and trust. For South Africa's construction industry, adopting a standardized approach to information management can help institutionalize these values, ensuring that project data supports long-term sustainability and resilience. Information management programs must therefore empower construction organizations to operate transparently, monitor project outcomes effectively, and maintain reliable records that support both legal compliance and stakeholder engagement. In recent years, reforms within South Africa's construction and public infrastructure sectors have emphasised the importance of transparency, accountability, and improved service delivery (Adekunle et al., 2022c). These reforms extend beyond financial management to include human resources, legal compliance, and record-keeping practices. Strengthening information management is central to these reforms, as it enables organisations to respond effectively to client needs, reduce inefficiencies, and deliver value across the project lifecycle.

In summary, the efficiency, accountability, and transparency of the construction industry in South Africa are significantly influenced by construction information management. Construction companies can increase project outcomes, stakeholder trust, and the sector's overall contribution to sustainable national development by implementing organised, dependable, and transparent information systems.

5.4 Building Information Modelling in South Africa

There is an expanding consciousness of building data demonstrating in South Africa among architects, with patterns and use designs mirroring the discoveries of different nations. Nonetheless, South African project workers are generally uninformed of the idea, with just 4% of workers for hire saying that they knew about the idea (Froise & Shakantu, 2014, p. 897). This fluctuates considerably with perceptions from different nations, like the USA, where most of the project workers had already embraced constructing data demonstrating (Bryed et al., 2012, p. 973). Not all the faculty in the South African construction industry are working in a BIM just climate; most experts are yet to embrace the utilization of BIM (Akintola et al., 2017, p. 1). At the point when this degree of reception is contrasted with reception rates in different nations and the dissemination of advancement hypothesis talked about by Kaminski (2011), it could be expected that compositional reception will begin moving into the late greater part stage, while project workers are entering the phase of consciousness of the development that recommends that reception of building data displaying rates should begin to increment in South Africa.

One might say that there are two attributes of building information modelling reception in South Africa, which are connecting autonomously (Froise & Shakantu, 2014: 896). The first is the reception of building information modelling by firms, which is a component of individual outcome. This offers many benefits for the creation of data. The subsequent trademark is the reception of the new working strategies related with building information modelling. Reception in different nations is happening on an industry-wide and public level, which is a component of public result (Harrison & Zack, 2012: 553). While reception of building information modelling on a hierarchical level is pursuing worldwide directions, this isn't seen at a public or modern level in South Africa (Adekunle et al., 2024c). The current antagonistic climate can't be tended to utilizing building information modelling in disengagement. Harris and Nawawi (2013: 5) additionally talk about how the reception of building data displaying occurs at two levels. For an innovation to be embraced at the public level, the public authority should command or control it. This outcomes in strategies for public guidelines and authorization acts, for example, making it obligatory for BIM use on open works projects in the event that it tends to be shown that there are benefits (Maconachie, 2012: 706).

Additionally, the execution of building information modelling to a great extent relies upon the readiness of the South African construction industry's stakeholders (Kekana et al., 2014: 109). The collaboration and venture of time and cash into achieving the abilities needed to convey building information modelling completely relies upon stakeholders (Alam, 2013: 103). Such speculations and endeavors won't squander as building information modelling will emphatically influence all gatherings included. The main part of building information modelling is that it permits the plan group overall, from the customer directly down to the providers, to team up better and work all the more precisely and productively (Azhar, 2011: 242). Because of the idea of building information modelling, working connections have become more strong and successful by ethicalness of making it feasible for partners to refresh their piece of work when one of the partners includes or disposes of some data the data set. This is conceivable in light of the fact that the entire virtual model of the structure, just as the data of the project, will be housed on exactly the same information base; this can also be known as a solitary task record (Tse et al., 86).

There are benefits of utilizing building information modelling in the South African construction industry all through all the project stages (Haron et al., 2010, p. 4). In the early plan and arranging stages, the contractor and architect are managed the cost of the chance of utilizing building information modelling devices, for example, ArchiCAD and Revit related to help their goals of creating the most ideal plan which is in accordance with the important rules of its planned area (Bryceson & Mackinnon, 2012, p. 514). For the cost experts, the general simplicity of removing the consolidating instruments, for example, Innovaya and Autodesk QTO, from the information extracted from the devices utilised by the architect. Additionally, the principle benefit of building information modelling has been distinguished to be further developed perception, further developed usefulness because of availability of information, and expanded coordination of reports (Haron et al., 2010: 5). This is mostly due to the multi-client joint effort capacities (Adekunle et al., 2024d). At the point when every one of the experts has taken care of business on their favored projects (BIM instruments), they then, at that point, add it to the information base of the project (Biao et al., 2006: 452). Furthermore, on the grounds that all of this is done electronically, it makes it promptly accessible consistently for different experts. One more advantage as distinguished is building information modelling capacity of implanting and connecting of imperative information like indicated materials, area of subtleties, and every one of the amounts needed for offering and assessing purposes (Gu & London, 2010: 989). The overall decrease of expenses has additionally been distinguished as an advantage (Du et al., 2018: 239). This is as time spent on the planning and coordination of the documentation, just as on the duplication of printed versions and the printing and replicating of documentation as it will be electronically open (He et al., 2008: 733). From the abovementioned, it tends to be presumed that building information modelling offers various advantages and benefits to various clients. The advantages of building information modelling range all through the different phases of a task, from the preparation through to upkeep of the finished project (Otasowie et al., 2023a).

5.5 Common Data Environment in South Africa's Construction Industry

A Common Data Environment (CDE) is a digital platform that centralises the collection, management, and dissemination of information across the lifecycle of construction projects (Otasowieet al., 2023b). Although there is a wealth of academic research globally on CDEs, scholarly investigations specifically focused on South Africa remain limited. One relevant study explores the critical success factors (CSFs) for implementing cloud-based data management, an enabler of CDE workflows, in the South African construction industry. This research, conducted among construction professionals and IT experts in Gauteng, identifies key CSFs such as awareness of cloud services, top management commitment, assurance of privacy, as well as clusters grouped under "Efficiency and flexibility," "Organizational and external," "Usability," and "Organizational scale" Adekunle et al., (2024c). This illustrates that organizational readiness, leadership, and secure, user-friendly systems are foundational to adopting CDE-like tools in the South African context (Mpingana et al., 2025).

Moreover, scholarly work on BIM adoption readiness in South Africa suggests that poor information management contributes to cost and time overruns, and that regulatory and mandatory initiatives may be necessary to accelerate implementation. A master's thesis investigating stakeholder perspectives points out that, although BIM is globally growing, South Africa lacks such mandates and requires tailored regulatory strategies to foster industry-wide uptake (Chimhundu, 2016). Another peer-reviewed conference paper examines BIM adoption within higher education institutions in South Africa. This study finds that there is significant resistance to integrating BIM into architectural, engineering, and construction curricula, largely due to a lack of government support, which has downstream implications for digital proficiency in the workforce (Armoed & Naidu, 2024). Without enough digital literacy training, the human capital necessary for effective CDE deployment remains underdeveloped.

Complementing these, although not specific to South Africa, an international journal article provides useful theoretical insights into CDE as a concept within BIM-enabled AEC (Architecture, Engineering, Construction) projects. It articulates that a CDE functions as a central repository enhancing collaboration by managing shared models, project data, and team communication. It further emphasizes that aligning business processes with BIM processes, managing data generation and consumption, and clarifying rights and responsibilities are essential for effective CDE implementation (Dolla et al., 2024).

Combining the insights from these academic sources, a cohesive narrative emerges: effective CDE adoption in South Africa hinges on organizational leadership, regulatory support, technical literacy, and strategic alignment of workflows to digital standards. Given the current lack of mandated BIM/CDE frameworks, investing in education and policy remains essential. As the master's thesis by Chimhundu (2016) argues, public sector leadership and formal initiative would provide impetus. At the same time, strengthening education pathways, as highlighted by Armoed and Naidu (2024), would contribute to building the necessary skills to operate within CDEs. Meanwhile, Adekunle et al. (2024e) (provide empirical guidance on what organizational factors are most vital to driving adoption, notably management commitment, usability, privacy assurance, and system flexibility.

Looking forward, South Africa could benefit from pilot programs in both industry and academia that demonstrate CDE value in real project settings, supported by policy incentives and integrated curricula. Leaders in construction firms could focus on the CSFs identified by Ikuabe et al. (2025a), while the public sector formulates initiatives along the recommendations of Tambwe et al., (2025). Simultaneously, academic institutions should address gaps in BIM pedagogy as highlighted by Ikuabe et al. (2025b), thereby preparing future professionals to engage with digital environments seamlessly.

5.6 State of the Construction Industry in India

The construction industry in India is one of the largest sectors of the economy and plays a pivotal role in infrastructure development, employment creation, and overall GDP growth. It contributes approximately 8–9% to the country's GDP and is the second-largest employer after agriculture, engaging more than 50 million people directly and indirectly (Kumar & Bansal, 2020). The sector is broadly divided into real estate construction, including residential, commercial, and industrial projects, and infrastructure construction, which encompasses roads, highways, railways, ports, airports, and energy-related facilities. Over the past two decades, India has witnessed rapid urbanisation, population growth, and increased investment in infrastructure, which have collectively driven the expansion of the construction industry. Government initiatives such as "Make in India," "Smart Cities Mission," and "Housing for All by 2022" have further accelerated growth in both public and private sector construction activities (Mishra & Jha, 2021).

A notable strength of the Indian construction industry is its role as a driver of urban transformation and modernisation. The rapid growth of metropolitan cities such as Delhi, Mumbai, Bengaluru, and Hyderabad has spurred demand for high-rise residential complexes, commercial hubs, and large-scale transport networks. The government's focus on mega projects such as Bharatmala (roadways), Sagarmala (ports), and the Dedicated Freight Corridor for railways indicates a strategic move toward developing integrated infrastructure that can enhance logistics and connectivity nationwide (Sharma & Sharma, 2019). Additionally, the growth of affordable housing initiatives under the Pradhan Mantri Awas Yojana has encouraged private sector participation in low-cost housing, further fueling construction activities across urban and semi-urban areas.

Despite these opportunities, the industry faces significant challenges and structural bottlenecks. One persistent issue is project delays and cost overruns, often caused by land acquisition hurdles, bureaucratic red tape, and inefficiencies in project management (Doloi et al., 2012). Studies show that a large percentage of infrastructure projects in India suffer from time and cost overruns due to poor planning, lack of coordination among stakeholders, and inadequate risk management. Furthermore, the sector is highly fragmented, with a large number of small and medium-sized contractors who often lack access to modern technologies and financial resources. This fragmentation contributes to inefficiency and reduces the ability to adopt innovative practices such as Building Information Modelling (BIM) and digital project management tools (Arora & Awasthi, 2020).

Another major concern is the labour-intensive nature of the industry and the dependence on unskilled or semi-skilled workers. Migrant labourers form the backbone of construction in India, but they often work under poor conditions with limited job security, safety, or training. The COVID-19 pandemic highlighted these vulnerabilities, as millions of migrant workers returned to their villages, causing a severe shortage of manpower and stalling projects across the country (Ghosh, 2021). Although the government has announced labour reforms and increased investment in skill development, the industry continues to grapple with low productivity and occupational health and safety issues.

In terms of sustainability, the Indian construction industry has increasingly been criticized for its environmental impact. Construction activities contribute significantly to carbon emissions, energy use, and waste generation. With India being one of the fastest urbanizing countries, the challenge of balancing development with sustainable practices is immense. Recent research points toward the growing adoption of green building certifications such as GRIHA and LEED, along with the use of renewable energy materials, energy-efficient designs, and sustainable urban planning models (Gupta, 2020). However, the uptake of sustainable practices is still limited to large developers, with smaller firms lagging behind due to higher costs and lack of awareness.

On the positive side, the sector is gradually witnessing a digital transformation. Technologies such as BIM, modular construction, prefabrication, drones, and Internet of Things (IoT) applications are slowly being integrated into projects. While adoption is currently at a nascent stage compared to developed economies, the push for "Digital India" and the government's emphasis on smart infrastructure are expected to accelerate the use of digital solutions in construction (Arora & Awasthi, 2020). The availability of foreign direct investment (FDI) in construction and real estate, coupled with international partnerships, is also expected to bring technological expertise into the sector.

The construction industry in India remains a critical driver of economic growth and development, with vast opportunities in housing, infrastructure, and smart city projects. However, structural inefficiencies, labour challenges, and environmental concerns continue to impede its full potential. To sustain growth, there is a need for stronger policy frameworks, streamlined project approval mechanisms, improved workforce training, and enhanced adoption of digital and sustainable practices. With government backing and private sector participation, the Indian construction industry is poised to play an even more significant role in shaping the country's future economy and urban landscape.

5.7 Construction Information Management in India

Construction Information Management (CIM) in India is an emerging field that focuses on the collection, storage, processing, and dissemination of project-related information throughout the construction lifecycle. With the rapid growth of the Indian construction sector, the need for structured information management has become critical to address challenges such as cost overruns, delays, and coordination failures (Doloi et al., 2012). Traditionally, Indian construction projects have relied on paper-based documentation and fragmented communication among stakeholders, which has often resulted in miscommunication, duplication of efforts, and project inefficiencies (Arora & Awasthi, 2020). The adoption of digital technologies, particularly Building Information Modelling (BIM), cloud-based project management platforms, and automated documentation systems, is gradually transforming the way information is managed in Indian construction projects.

Academic research highlights that the adoption of CIM practices in India remains nascent, with significant variation across project types and organizational size. Large infrastructure and urban development projects are more likely to adopt digital CIM tools due to higher budgets, complex coordination needs, and external stakeholder requirements, whereas small- and medium-scale construction projects often continue to rely on conventional, less structured methods (Mishra & Jha, 2021). The effectiveness of CIM depends on a combination of organizational factors such as management support, technical skill availability, and standardization of procedures. For example, Doloi et al. (2012) emphasize that poor information flow is one of the primary causes of time and cost overruns in Indian construction, suggesting that implementing structured information management systems could significantly improve project performance.

BIM-based CIM is increasingly recognized as a strategic enabler for project coordination and decision-making in India. Studies indicate that BIM facilitates real-time collaboration among architects, engineers, contractors, and facility managers, integrating design, cost estimation, scheduling, and procurement information into a single platform (Arora & Awasthi, 2020). However, BIM adoption faces challenges, including high initial investment costs, lack of trained professionals, and resistance to change from traditional project management practices. A survey of Indian construction firms reveals that organizations with prior exposure to BIM and digital project management tools report improvements in productivity, reduced errors, and enhanced communication (Kumar & Bansal, 2020).

Another critical aspect of CIM in India is data standardization and governance. Unlike countries with established national BIM mandates, India lacks comprehensive regulatory frameworks for information management, leading to inconsistent standards and practices across projects (Sharma & Sharma, 2019). Researchers suggest that the development of national guidelines and frameworks for information exchange, document management, and data security could significantly enhance the adoption of structured CIM practices. Moreover, the integration of cloud-based project management systems and mobile technology is facilitating access to project data on-site, enabling faster decision-making and improved coordination among geographically dispersed teams (Mishra & Jha, 2021).

The COVID-19 pandemic has further highlighted the importance of robust CIM practices in India. Remote work, social distancing, and interruptions in labour availability necessitated digital solutions for monitoring project progress, sharing information, and ensuring continuity of operations (Ghosh, 2021). This shift has accelerated the adoption of cloud-based tools and mobile applications for information management, signalling a growing awareness of the value of effective CIM in mitigating risks and maintaining project efficiency.

Construction Information Management in India is evolving from traditional paper-based methods toward digital, integrated solutions driven by BIM, cloud technologies, and mobile platforms. While adoption is still uneven and faces challenges such as lack of standardization, skilled personnel, and resistance to change, academic studies indicate that structured CIM can significantly improve project coordination, reduce errors, and enhance overall efficiency. Strengthening regulatory frameworks, investing in professional training, and promoting digital tools are key strategies for advancing CIM in the Indian construction sector.

5.8 Building Information Modelling in India

Building Information Modelling (BIM) in India has emerged as a transformative approach to managing construction projects by integrating design, planning, construction, and facility management into a collaborative digital environment. BIM enables the creation and management of 3D models enriched with information regarding materials, scheduling, cost, and operational details. In India, the adoption of BIM is still in its nascent stage, largely concentrated in large-scale infrastructure, metro projects, and multinational construction firms (Arora & Awasthi, 2020). Government initiatives such as the "Smart Cities Mission" and major metro rail projects have acted as catalysts for BIM adoption by emphasising the need for efficiency, cost control, and data-driven decision-making (Sharma & Sharma, 2019).

Sharma and Sharma (2019) indicate that BIM adoption in India faces both technical and organisational challenges. Technically, limited availability of skilled professionals, high implementation costs, and the lack of standardised processes across firms hinder widespread adoption (Kumar & Bansal, 2020). Organizationally, resistance to change, fragmented workflows, and low awareness among small- and medium-sized contractors slow the integration of BIM into conventional project management practices (Arora & Awasthi, 2020). Nevertheless, projects that successfully implement BIM report significant benefits, including improved project coordination, reduced rework, enhanced visualisation, and more accurate cost and time estimates.

BIM is particularly valuable in large infrastructure and urban development projects in India. For instance, metro rail projects in Delhi, Bengaluru, and Mumbai have leveraged BIM to integrate design and construction data, improving clash detection, procurement planning, and coordination among multiple stakeholders (Sharma & Sharma, 2019). Similarly, the adoption of BIM in high-rise residential and commercial buildings enables visualization of complex designs and facilitates efficient communication between architects, engineers, and contractors. Research also highlights that BIM integration with 4D scheduling and 5D cost estimation provides a holistic project management framework that reduces delays and budget overruns (Mishra & Jha, 2021).

Despite the demonstrated advantages, policy and regulatory frameworks supporting BIM adoption in India remain underdeveloped. Unlike countries such as the UK, Singapore, or the USA, India has not yet implemented national mandates or standards for BIM implementation. Consequently, adoption is often voluntary, driven by project complexity, client demand, or multinational collaborations (Arora & Awasthi, 2020). Scholars emphasize the need for governmental incentives, standardized guidelines, and professional training programs to accelerate BIM adoption across public and private sector projects (Chandrasekaran & Ganesan, 2021).

The role of digital skills and education is also critical in advancing BIM adoption. Universities and technical institutes are gradually incorporating BIM-related courses into architecture, civil engineering, and construction management curricula to prepare graduates for industry-ready digital workflows. Studies suggest that training programs focusing on software proficiency, project coordination, and BIM standards can significantly reduce adoption barriers and increase workforce competency (Arora & Awasthi, 2020).

Building Information Modelling in India represents a promising but evolving frontier in construction management. While current adoption is limited to large-scale projects and progressive firms, BIM has the potential to enhance project efficiency, cost control, and collaboration across the construction lifecycle. For India to fully realize the benefits of BIM, there is a need for policy support, skill development, standardized guidelines, and broader industry awareness. As governmental initiatives, private sector engagement, and academic training converge, BIM adoption in India is likely to expand, positioning the country to meet the challenges of rapid urbanization and infrastructure development more effectively.

5.9 Common Data Environment in India's Construction Industry

A Common Data Environment (CDE) in the Indian construction industry refers to a collaborative digital platform where all project-related information—such as drawings, models, schedules, cost data, and documentation, is stored, managed, and shared among stakeholders. CDEs are integral to Building Information Modelling (BIM) processes and serve as a single source of truth, enabling real-time access, version control, and audit trails (Arora & Awasthi, 2020). In India, the adoption of CDEs is still in its early stages, predominantly in large-scale infrastructure, metro projects, and high-end commercial developments. Government-led initiatives, such as the Smart Cities Mission and metro rail expansions, have encouraged the integration of digital project delivery systems, providing a platform for CDE deployment (Sharma & Sharma, 2019).

Ade et al. (2024) indicates that fragmented project workflows and reliance on traditional paper-based documentation remain major barriers to effective CDE implementation in India. Many construction firms, particularly small and medium enterprises, lack the technical expertise and resources necessary to adopt standardized digital environments (Kumar & Bansal, 2020). Furthermore, inconsistent information exchange protocols, absence of regulatory mandates, and low awareness of CDE benefits contribute to uneven adoption across the industry (Chandrasekaran & Ganesan, 2021). These challenges result in inefficiencies, duplicated effort, and poor coordination among project stakeholders, which can ultimately lead to cost and time overruns.

Despite these barriers, studies have shown that CDEs offer significant advantages when integrated into Indian construction projects. By centralizing design models, cost data, schedules, and project documentation, CDEs facilitate enhanced collaboration among architects, engineers, contractors, and clients (Arora & Awasthi, 2020). They also support 4D scheduling and 5D cost management, allowing for better visualization of construction sequences and more accurate budgeting. In addition, CDEs improve information traceability, risk management, and decision-making, aligning with global BIM standards such as ISO 19650, which emphasizes structured data management throughout a project's lifecycle (Mishra & Jha, 2021).

The integration of CDEs in India is also closely linked with the growth of digital construction technologies. Tools such as cloud-based project management platforms, mobile information systems, and BIM-enabled collaboration software are increasingly being adopted in urban infrastructure and large-scale projects. These technologies provide remote access to project data, facilitate coordination among geographically dispersed teams, and reduce the reliance on physical documentation (Chandrasekaran & Ganesan, 2021). Additionally, pilot projects in metro rail construction and large commercial developments have demonstrated improvements in workflow efficiency, error reduction, and timely project delivery when CDE principles are applied effectively.

However, the future of CDE adoption in India depends on several critical factors. These include the development of standardised protocols for information management, investment in workforce training, and increased awareness of the strategic benefits of CDEs. Policy support from government agencies, particularly in public infrastructure projects, can accelerate adoption by incentivising compliance with digital information management practices (Sharma & Sharma, 2019). Furthermore, academic institutions and professional training programs can enhance the technical skills of engineers, architects, and construction managers, enabling them to effectively utilise CDE platforms.

The Indian construction industry is gradually moving toward digitally enabled project management, with CDEs playing a central role in this transformation. While adoption is currently limited to large-scale and complex projects, the integration of CDEs promises significant improvements in project coordination, cost control, and efficiency. Addressing challenges related to standardisation, training, and policy will be essential for expanding CDE use across India's broader construction sector. As India continues its rapid urbanization and infrastructure development, CDEs are poised to become a critical enabler of modern, collaborative, and data-driven construction practices.

5.10 State of the Construction Industry in Egypt

The construction industry in Egypt plays a pivotal role in the country's economic development, contributing significantly to gross domestic product (GDP), employment creation, and infrastructure modernization. According to the World Bank, construction has consistently accounted for between 5 and 6% of Egypt's GDP, with a major share coming from housing, transport infrastructure, and large-scale public projects (El-adaway et al., 2020). The sector has experienced rapid growth over the past decade, driven by urban expansion, population growth, and government investment in national megaprojects. The Egyptian government has prioritized construction as part of its long-term development strategy, Egypt Vision 2030, which emphasizes infrastructure improvement, new urban communities, and industrial development (Attia & Ibrahim, 2018).

A key driver of the Egyptian construction industry is the government's ambitious megaproject agenda. Projects such as the construction of the New Administrative Capital, the Suez Canal Area Development, extensive road networks, new power plants, and the expansion of housing programs have positioned Egypt as one of Africa's largest construction markets (Marzouk & El Kherbawy, 2019). The New Administrative Capital alone, envisioned as a smart and sustainable city, reflects Egypt's strategy to decongest Cairo and stimulate economic growth through construction-led urban development. These projects are largely financed through public-private partnerships (PPPs), foreign investments, and international funding, particularly from China and Gulf countries, which have strengthened Egypt's infrastructure base (Osman et al., 2021).

Despite this growth, the construction sector in Egypt faces numerous challenges. One of the most critical issues is the high degree of fragmentation within the industry, where traditional practices dominate, and collaboration between stakeholders is often weak (Hegazy & El-adaway, 2017). Cost overruns, project delays, and quality concerns are recurring challenges, often linked to inadequate project management practices, fluctuating material costs, and limited adoption of digital construction technologies such as Building Information Modelling (BIM). Additionally, the reliance on manual processes and insufficient skilled labor remains a barrier to productivity improvement (El-adaway et al., 2020).

Another challenge relates to the industry's regulatory environment and financing mechanisms. While PPPs have been promoted as a solution for funding large-scale infrastructure, issues of contractual disputes, risk allocation, and institutional inefficiencies have hindered their full potential (Osman et al., 2021). Moreover, inflation, currency fluctuations, and external shocks, such as the COVID-19 pandemic, have put pressure on construction firms, increasing the cost of materials and disrupting supply chains (Marzouk & El Kherbawy, 2019). Environmental sustainability also poses a major concern, as rapid construction expansion has raised issues related to energy consumption, waste generation, and carbon emissions.

On the other hand, there are promising trends shaping the future of the construction industry in Egypt. The government and private sector are increasingly showing interest in adopting sustainable construction practices, such as green building technologies, energy-efficient design, and renewable energy integration (Attia & Ibrahim, 2018). Furthermore, there is growing awareness of the need to digitize construction through BIM, common data environments (CDEs), and project management software to reduce inefficiencies and improve collaboration. Universities and professional organizations are also working to enhance technical education and research in construction management, which is expected to improve human resource capacity in the long term (Hegazy & El-adaway, 2017).

The Egyptian construction industry stands at the crossroads of rapid expansion and systemic challenges. It is a key contributor to national economic growth and urban development, particularly through government-led megaprojects and infrastructure modernization. However, structural challenges such as fragmentation, limited digital adoption, and financing constraints continue to hinder efficiency. With targeted reforms, capacity-building initiatives, and increased adoption of modern technologies, Egypt's construction industry has the potential to not only sustain its growth trajectory but also align with international standards of efficiency, sustainability, and resilience.

5.11 Construction Information Management in Egypt

Construction Information Management (CIM) is increasingly recognized as a critical factor for improving efficiency, collaboration, and decision-making in the Egyptian construction industry. Traditionally, the sector in Egypt has relied heavily on paper-based records, fragmented communication, and manual coordination between stakeholders. This has often led to information silos, miscommunication, and inefficiencies that result in delays, cost overruns, and disputes (Hegazy & El-adaway, 2017). With the rise of complex megaprojects such as the New Administrative Capital, metro expansions, and large-scale housing schemes, the need for structured information management systems has become urgent.

A major issue facing information management in Egyptian construction is the lack of integration across the project lifecycle. Project stakeholders—including contractors, consultants, and government agencies—often work independently using disparate documentation systems. This fragmentation makes it difficult to track project progress, share data transparently, and ensure accountability (El-adaway et al., 2020). Construction contracts typically lack clear provisions for digital information exchange, leading to disputes over responsibilities and data ownership. As a result, projects suffer from duplicated work, errors in documentation, and inconsistent record-keeping.

Recent studies highlight that the adoption of digital tools such as Building Information Modelling (BIM) and Common Data Environments (CDEs) remains at an early stage in Egypt. Although some large contractors and international consultants have begun implementing BIM for design coordination, its use is not widespread across the industry (Marzouk & El Kherbawy, 2019). The absence of national BIM mandates and standardized information protocols limits the effectiveness of such digital initiatives. In contrast, developed countries have demonstrated that structured CIM supported by BIM and CDEs significantly improves productivity, reduces errors, and enhances collaboration across disciplines. Egypt's challenge, therefore, is to establish a regulatory and technological framework that fosters digital information management practices.

In addition, human resource capacity and training remain critical barriers to effective CIM in Egypt. Many construction professionals have limited exposure to advanced digital tools, and project teams often lack the skills needed to create, manage, and exchange structured digital information (Osman et al., 2021). Universities and professional organizations are gradually integrating BIM and digital project management into their curricula, but the pace of change has been slow. This creates a gap between the industry's needs for modern information management practices and the available workforce competencies.

Furthermore, cultural and organizational barriers hinder the adoption of robust information management practices. Many stakeholders are resistant to change due to a reliance on traditional methods and a lack of trust in digital collaboration platforms. Contractual relationships in Egypt are often adversarial, which discourages data sharing and transparency (Hegazy & El-adaway, 2017). Without clear frameworks for data governance, intellectual property rights, and risk allocation, stakeholders remain hesitant to fully embrace digital information management systems.

Despite these challenges, the future of construction information management in Egypt appears promising. The government's ambitious infrastructure agenda under Egypt Vision 2030 places emphasis on sustainable urban development and modernization, which will require digital integration in construction processes (Attia & Ibrahim, 2018). International collaboration, foreign investment, and exposure to global best practices are gradually pushing Egyptian firms to adopt more advanced information management approaches. Increasingly, public-private partnerships (PPPs) and international consultants are introducing digital platforms for project monitoring, which may accelerate the adoption of CIM across the sector.

Construction information management in Egypt is still in a transitional stage, shifting from traditional fragmented approaches to digital and collaborative systems. The industry faces structural, technical, and cultural challenges, including limited adoption of BIM, a lack of standardised protocols, and skill shortages. However, with the continued push for infrastructure development and exposure to international best practices, Egypt has the potential to significantly enhance its construction information management practices, aligning them with global trends to improve project efficiency, sustainability, and stakeholder collaboration.

5.12 Building Information Modelling in Egypt

Building Information Modelling (BIM) has emerged as a transformative tool in the global construction industry, offering advanced capabilities for project visualization, coordination, and information management across the lifecycle of buildings and infrastructure. In Egypt, the adoption of BIM is steadily gaining traction, although its implementation remains at an early stage compared to more developed countries. The push towards BIM in Egypt has largely been driven by the complexity of recent megaprojects, such as the New Administrative Capital, large-scale housing programs, and transport infrastructure projects. These initiatives have highlighted the limitations of traditional design and project management approaches, thereby creating opportunities for BIM integration (Marzouk & El Kherbawy, 2019).

The Egyptian construction industry faces several structural and operational challenges that make BIM adoption both necessary and difficult. On the one hand, BIM has the potential to enhance coordination among project stakeholders, reduce errors in design, and mitigate cost and time overruns—issues that are prevalent in Egyptian construction projects (Hegazy & El-adaway, 2017). On the other hand, barriers such as the lack of government mandates, absence of standardized guidelines, and limited awareness among industry professionals have slowed down its widespread application (El-adaway et al., 2020). Unlike countries such as the UK and Singapore, which have introduced national BIM strategies and regulatory frameworks, Egypt has yet to establish a comprehensive policy for BIM implementation.

The current use of BIM in Egypt is largely concentrated among multinational firms, large contractors, and consultants who operate on high-profile projects. These organizations often adopt BIM to meet international standards or to fulfill contractual obligations with foreign partners. For smaller firms, however, the cost of BIM software, the need for specialised training, and the absence of clear contractual frameworks for BIM deliverables represent significant obstacles (Osman et al., 2021). Furthermore, the collaborative culture required for BIM implementation often clashes with the traditionally adversarial and fragmented nature of contractual relationships in Egypt's construction industry.

El-Anwar et al. (2021) highlight that BIM adoption in Egypt is also constrained by capacity and skills shortages. Universities and training institutions have only recently begun incorporating BIM into their curricula, which means that the majority of the workforce is not adequately trained in BIM processes and tools (Soliman et al., 2018). This gap between academic preparation and industry demand creates a bottleneck for BIM diffusion across the sector. Moreover, limited investment in continuous professional development programs exacerbates this challenge.

Despite these barriers, the potential benefits of BIM in Egypt are substantial. BIM can provide a single source of truth for project data, enabling better decision-making, improved risk management, and enhanced collaboration between stakeholders. In the context of Egypt's rapid urbanization and the government's ambitious Egypt Vision 2030, which prioritizes sustainable urban development, BIM could play a pivotal role in achieving efficiency and sustainability goals (Attia & Ibrahim, 2018). Studies also suggest that BIM could support Egypt in addressing challenges related to construction waste, energy efficiency, and lifecycle asset management—key issues in its growing urban infrastructure (Marzouk & Hisham, 2019). Furthermore, the integration of BIM with Common Data Environments (CDEs) and emerging technologies such as Geographic Information Systems (GIS), Internet of Things (IoT), and Artificial Intelligence (AI) offers promising opportunities for Egypt. These integrations could support more advanced applications such as digital twins, predictive maintenance, and smart city planning, which align with Egypt's modernisation agenda (El-Anwar et al., 2021).

To sum up, BIM in Egypt is at a transitional stage, with notable progress in high-profile projects but limited penetration across the broader industry. The key challenges include lack of regulatory frameworks, high implementation costs, skills shortages, and cultural resistance to collaborative practices. Nevertheless, with the increasing demand for efficient and sustainable infrastructure, along with exposure to international practices, BIM adoption in Egypt is expected to grow. For Egypt to fully harness the potential of BIM, it will require national-level strategies, standardized guidelines, and capacity-building initiatives that can create an enabling environment for digital transformation in construction.

5.13 Common Data Environment in Egypt's Construction Industry

The construction industry in Egypt, like many developing countries, is undergoing a significant transformation driven by digitalisation and the adoption of advanced technologies such as Building Information Modelling (BIM). At the heart of BIM-enabled collaboration lies the Common Data Environment (CDE)—a centralized digital repository designed to store, manage, and disseminate project information across the lifecycle of a project. A CDE facilitates seamless collaboration among multiple stakeholders by providing a single source of truth, thereby reducing errors, duplication, and inefficiencies that have traditionally plagued construction projects. In the Egyptian context, where projects are often large-scale, complex, and involve numerous actors, the implementation of CDE has the potential to address chronic issues such as project delays, cost overruns, and fragmented communication (Soliman et al., 2018).

CDE adoption in Egypt is closely tied to the gradual diffusion of BIM practices in the country. While BIM has gained visibility in megaprojects such as the New Administrative Capital and large infrastructure schemes, the supporting infrastructure of CDEs is still in its infancy. Unlike in countries such as the UK, where BIM Level 2 mandates necessitate the use of CDEs, Egypt lacks a formal regulatory framework that enforces their deployment (El-adaway et al., 2020). Consequently, the use of CDEs is often project-specific and largely driven by international firms or large Egyptian contractors engaged in projects requiring compliance with international standards. This fragmented adoption creates disparities in efficiency, as smaller contractors and consultants are often excluded from the digital collaboration process due to limited resources or technological capacity.

The benefits of adopting CDEs in Egypt's construction industry are substantial. A CDE provides a platform for real-time information exchange, version control, and coordination of models and documents, which are particularly critical in Egypt's highly dynamic construction environment. Studies emphasize that the absence of effective communication and integrated information systems is one of the main contributors to rework, disputes, and claims in Egyptian projects (Hegazy & El-adaway, 2017). By ensuring that all stakeholders work with the most up-to-date information, CDEs could significantly reduce such inefficiencies. Moreover, the adoption of CDEs aligns with Egypt's long-term goals of modernizing its infrastructure and construction processes under Egypt Vision 2030, which emphasizes digital transformation and sustainability (Attia & Ibrahim, 2018).

Despite these potential benefits, several barriers hinder the widespread implementation of CDEs in Egypt. One major challenge is the lack of technical knowledge and training among industry professionals. While universities and training institutions have started incorporating BIM education into their curricula, specific training on CDE usage remains limited (Soliman et al., 2018). Furthermore, the high cost of acquiring, operating, and maintaining CDE platforms is prohibitive for many small and medium-sized firms, which dominate Egypt's construction market. The absence of local CDE providers further complicates adoption, as firms are often reliant on expensive international solutions such as Autodesk BIM 360 or Bentley ProjectWise (Marzouk & El Kherbawy, 2019).

Cultural and organizational barriers also play a significant role in limiting CDE uptake. Egypt's construction industry has historically been characterized by fragmented supply chains, adversarial contractual relationships, and limited collaboration between stakeholders (Osman et al., 2021). The collaborative ethos required for successful CDE implementation often clashes with entrenched practices that prioritize siloed operations and individual accountability over shared responsibility. Overcoming these cultural barriers requires a significant shift in project management approaches, supported by training, awareness campaigns, and strong leadership from both public and private sector actors.

Looking ahead, the role of CDEs in Egypt's construction industry is expected to expand, especially as the government and private sector increasingly recognise the importance of digitalisation in achieving efficiency and sustainability. Research suggests that integrating CDEs with emerging technologies such as the Internet of Things (IoT), Geographic Information Systems (GIS), and cloud computing could further enhance their value, enabling advanced applications such as digital twins and smart infrastructure management (El-Anwar et al., 2021). For example, linking CDEs to real-time data streams could improve asset management and predictive maintenance for large infrastructure projects, which are central to Egypt's development agenda.

In a nutshell, while the use of Common Data Environments in Egypt's construction industry is still limited, their potential to revolutionize project delivery is significant. The current state of adoption is largely driven by multinational firms and megaprojects, but widespread integration is hindered by challenges such as cost, lack of expertise, absence of regulatory mandates, and cultural resistance. To unlock the full benefits of CDEs, Egypt will need to establish national guidelines, promote local technological solutions, and invest in capacity building for construction professionals. With these measures, CDEs could play a pivotal role in transforming Egypt's construction industry into a more collaborative, efficient, and sustainable sector.

5.14 State of the Construction Industry in Brazil

The construction industry in Brazil plays a vital role in the country's economic and social development, contributing significantly to GDP, employment, and infrastructure provision. According to industry estimates, construction accounts for approximately 6–8% of Brazil's GDP and is among the largest employers, with millions of jobs created across formal and informal sectors (Sanches & Vieira, 2018). The sector encompasses a wide range of activities including residential and commercial building, infrastructure development, and industrial projects. Given Brazil's rapid urbanization, population growth, and infrastructural demands, the construction industry remains a central pillar in supporting sustainable development and improving quality of life. However, despite its importance, the industry faces persistent challenges relating to productivity, project performance, corruption, and macroeconomic instability.

Historically, Brazil's construction industry has been characterized by cyclical fluctuations closely tied to the country's broader economic conditions. Periods of rapid growth, such as during preparations for the 2014 FIFA World Cup and 2016 Olympic Games, were followed by steep downturns due to political and economic crises (Mattos, 2016). During the recession of 2014–2016, the construction sector contracted sharply as government investments in infrastructure declined, credit availability tightened, and several major firms became embroiled in corruption scandals, particularly the Lava Jato (Car Wash) investigation, which exposed systemic bribery and malpractice in large-scale construction contracts (Lima & da Silva, 2019). This crisis undermined investor confidence and reduced construction activity, highlighting structural weaknesses in governance and transparency within the sector.

Despite these setbacks, recent years have witnessed a slow but steady recovery. Government infrastructure programs, such as the Programa de Parcerias de Investimentos (PPI—Investment Partnership Program), have aimed to attract private investment into roads, ports, energy, and housing (Peci et al., 2020). Public-private partnerships (PPPs) have emerged as a critical mechanism for financing large-scale infrastructure, given fiscal limitations faced by the Brazilian government. The housing sector, in particular, has seen sustained growth due to initiatives like Minha Casa,

Minha Vida (My House, My Life), a social housing program that has provided affordable housing for millions of low-income families (Silva et al., 2018). These programs have not only stimulated construction activity but also sought to address Brazil's long-standing housing deficit, which remains one of the largest in Latin America.

Technological innovation and digitalization are increasingly shaping the construction industry in Brazil. Building Information Modelling (BIM) has gained attention as a tool to improve project coordination, reduce waste, and enhance efficiency. In 2018, the Brazilian government launched the Estratégia BIM BR (BIM BR Strategy), a national policy to foster BIM adoption across public projects (Costa & Furlani, 2020). This initiative aims to modernize construction practices, improve transparency in procurement, and align Brazil with global digital construction trends. Furthermore, sustainability has become an emerging priority, with increasing adoption of green building certifications such as LEED (Leadership in Energy and Environmental Design) and AQUA-HQE, reflecting growing awareness of environmental and social responsibilities (Amaral et al., 2019). However, the diffusion of these innovations is uneven, with larger firms adopting digital tools and sustainable practices more readily than small and medium enterprises (SMEs), which often lack resources and technical capacity.

Nevertheless, the Brazilian construction industry continues to grapple with structural challenges. Low productivity levels, fragmented supply chains, and poor project management practices contribute to delays, cost overruns, and inefficiencies (Souza & Kagioglou, 2019). Informality in the labor market remains a significant issue, as a large proportion of construction workers operate without formal contracts, social protection, or training, which undermines quality and safety. Occupational health and safety standards are often poorly enforced, leading to a high incidence of workplace accidents compared to global benchmarks (Oliveira et al., 2021). Corruption and governance issues also persist, particularly in public procurement processes, though anti-corruption reforms and judicial oversight are beginning to strengthen accountability.

Looking forward, the prospects of the Brazilian construction industry depend heavily on political stability, macroeconomic growth, and the government's ability to sustain infrastructure investments. Brazil faces significant infrastructure gaps, particularly in transportation, sanitation, and energy, which present opportunities for long-term construction growth. Digital transformation and sustainability will likely play increasingly important roles, with BIM adoption, modular construction, and smart technologies expected to reshape practices. However, realizing these opportunities will require overcoming barriers related to training, financing, and institutional reform.

The construction industry in Brazil remains a cornerstone of national development but continues to face complex challenges. While it has made strides in modernization, housing provision, and infrastructure development, issues such as corruption, productivity inefficiencies, and informality remain significant obstacles. Through strategic reforms, stronger governance, and greater integration of digital and sustainable practices, Brazil's construction industry has the potential to become more resilient, efficient, and globally competitive in the coming decades.

5.15 Construction Information Management in Brazil

Construction Information Management (CIM) has emerged as a critical aspect of the Brazilian construction industry, particularly in the context of increasing project complexity, urban development demands, and the drive for efficiency and transparency. CIM refers to the systematic processes of generating, managing, sharing, and storing information across a construction project's lifecycle to improve decision-making, coordination, and outcomes (Succar, 2009). In Brazil, the management of construction information is undergoing a significant transformation, driven by digital technologies such as Building Information Modelling (BIM), government initiatives, and the need to address longstanding inefficiencies in the industry.

Traditionally, the Brazilian construction sector has been characterized by fragmented information flows, reliance on paper-based documentation, and poor communication among stakeholders. This has often led to challenges such as delays, cost overruns, and disputes in project delivery (Costa & Furlani, 2020). A common problem has been the lack of integration between design, engineering, and construction phases, resulting in information silos and duplication of work. Furthermore, the high level of informality in the labor market and the dominance of small and medium-sized enterprises (SMEs) have limited the adoption of formalized information management practices (Souza & Kagioglou, 2019). As a result, construction information in Brazil has historically been underutilized as a strategic resource, with significant consequences for project performance and industry productivity.

In recent years, however, there has been a concerted push to modernize construction information management practices in Brazil. A key driver of this change is the national BIM mandate, known as the Estratégia BIM BR, launched in 2018 by the Brazilian government. This strategy aims to promote BIM adoption in public projects as a means of standardizing construction information, improving interoperability, and enhancing the transparency of procurement and contract management (Antunes & Scheer, 2020). Under the phased implementation plan, BIM use became mandatory in federal public tenders for certain types of projects starting in 2021, with broader requirements scheduled for subsequent years. This policy framework has created momentum for the digitalization of information management processes across both public and private sectors.

The adoption of BIM and related digital tools has fundamentally reshaped how construction information is generated and shared in Brazil. BIM platforms enable the creation of digital models that integrate architectural, structural, and systems information into a unified database, facilitating collaboration among stakeholders (Sacks et al., 2018). This integrated approach reduces inconsistencies, enables clash detection, and supports life-cycle asset management. In Brazil, large firms and multinational contractors have been at the forefront of BIM adoption, while SMEs face challenges due to limited technical expertise, high software costs, and resistance to organizational change (Barison & Santos, 2011). Nevertheless, professional training programs and academic research initiatives have begun to bridge these gaps, equipping professionals with the necessary skills to engage with digital information management practices.

Information management in Brazil's construction industry has also been influenced by broader trends in sustainability and smart cities. For example, CIM systems are increasingly being used to collect and analyze data on energy efficiency, environmental performance, and life-cycle costing (Amaral et al., 2019). The integration of Geographic Information Systems (GIS) with BIM has been explored in urban planning and infrastructure projects, supporting better decision-making and resource allocation in complex urban environments (Vazquez-Corral et al., 2020). These developments illustrate the growing recognition of construction information as a strategic resource not only for project delivery but also for sustainable urban development.

Despite these advances, significant challenges remain in institutionalizing effective information management practices in Brazil. Key barriers include the lack of standardized protocols for data exchange, limited enforcement of BIM guidelines across regions, and disparities in adoption between large firms and smaller contractors (Costa & Furlani, 2020). Cultural resistance to change also persists, as many stakeholders continue to rely on traditional documentation methods and are hesitant to invest in digital solutions. Moreover, the uneven availability of skilled professionals in information management presents a bottleneck for widespread adoption. Addressing these challenges will require a combination of policy enforcement, industry training, and investment in affordable digital solutions for SMEs.

Therefore, it is important to note that construction information management in Brazil is at a transitional stage, moving from fragmented, paper-based systems toward integrated, digitalised platforms enabled by BIM and related technologies. While government policies such as Estratégia BIM BR have accelerated this shift, the industry still faces structural and cultural challenges that limit the pace of adoption. With continued investment in training, standardisation, and digital infrastructure, CIM has the potential to significantly improve productivity, transparency, and sustainability in Brazil's construction sector, thereby aligning it with global best practices.

5.16 Building Information Modelling in Brazil

Building Information Modelling (BIM) has emerged as a transformative force in the global construction industry, and Brazil is no exception. BIM refers to the digital representation of a building's physical and functional characteristics, providing a collaborative platform for stakeholders throughout a project's lifecycle (Succar, 2009). In Brazil, BIM has gained momentum in recent years due to government initiatives, the need for improved efficiency, and increasing alignment with global construction practices. Despite persistent challenges such as fragmented industry practices, limited technical expertise, and high implementation costs, BIM adoption is gradually reshaping Brazil's construction landscape.

Historically, the Brazilian construction sector has been characterized by inefficiencies, project delays, and budget overruns, largely due to fragmented processes and the reliance on traditional 2D design methods (Souza & Kagioglou, 2019). This fragmentation has often created communication gaps between architects, engineers, contractors, and clients, resulting in information silos and costly rework. Against this backdrop, BIM is seen as a promising solution for integrating multidisciplinary information and fostering collaboration. By centralizing design, construction, and operational data, BIM has the potential to mitigate inefficiencies and enhance the competitiveness of the Brazilian construction industry in both domestic and international markets.

The turning point for BIM adoption in Brazil came with the launch of the Estratégia BIM BR (BIM BR Strategy) in 2018. This federal initiative set out a phased roadmap for mandatory BIM use in public projects, with the first phase requiring its application in federal tenders for certain building projects starting in 2021 (Antunes & Scheer, 2020). The strategy aims to standardise construction information, increase transparency in public procurement, and modernise the sector by promoting digital transformation. By mandating BIM in public contracts, the government is not only encouraging its adoption but also setting a precedent for private sector stakeholders to align with these practices.

The adoption of BIM in Brazil has been uneven across the industry. Large construction companies and multinational firms operating in Brazil have been at the forefront of implementation, investing in advanced technologies, training, and software (Costa & Furlani, 2020). These organisations have reported benefits such as improved design coordination, reduced errors, and enhanced project visualisation. However, small and medium-sized enterprises (SMEs), which dominate Brazil's construction market, often struggle with the financial and technical demands of BIM implementation (Barison & Santos, 2011). High software costs, lack of skilled professionals, and cultural resistance to change continue to pose significant barriers to widespread adoption.

In the academic and professional training spheres, Brazil has made notable progress in equipping future construction professionals with BIM skills. Universities and technical institutions have increasingly incorporated BIM into their curricula, fostering a new generation of architects, engineers, and construction managers who are proficient in digital construction methods (Sacks et al., 2018). Research collaborations between academia and industry are also expanding, focusing on issues such as BIM-GIS integration, sustainability, and interoperability (Vazquez-Corral et al., 2020). These efforts are crucial for overcoming the human capital challenges that hinder BIM's full adoption.

BIM is also being recognized in Brazil for its role in advancing sustainability and smart construction practices. For example, BIM has been applied in projects focused on energy efficiency, life-cycle costing, and environmental performance monitoring (Amaral et al., 2019). Its ability to simulate building performance and facilitate data-driven decision-making supports Brazil's broader goals of sustainable urban development and infrastructure modernization. Moreover, the integration of BIM with emerging technologies such as the Internet of Things (IoT), cloud platforms, and digital twins presents new opportunities for enhancing asset management and operational efficiency in the Brazilian context.

Despite these advances, challenges remain in fully institutionalizing BIM in Brazil. The lack of standardized protocols for data sharing and interoperability among different platforms creates inefficiencies. Furthermore, the pace of adoption is hindered by regional disparities, as BIM is more widely used in metropolitan areas compared to smaller cities. Another persistent issue is the uneven enforcement of the BIM mandate, as some public agencies lack the technical capacity to implement and oversee BIM requirements effectively (Costa & Furlani, 2020). Addressing these barriers will require stronger industry collaboration, government oversight, and incentives for SMEs to adopt digital tools.

Consequently, BIM in Brazil represents both a challenge and an opportunity. While the industry continues to grapple with financial, technical, and cultural obstacles, the combination of government mandates, academic initiatives, and industry innovation is steadily driving progress. If these efforts are sustained and scaled, BIM has the potential to significantly improve project outcomes, reduce inefficiencies, and align Brazil's construction sector with international best practices. The ongoing digital transformation of construction in Brazil through BIM adoption will be instrumental in shaping a more efficient, transparent, and sustainable industry.

5.17 Common Data Environment in Brazil's Construction Industry

The construction industry in Brazil, much like in other parts of the world, has historically faced significant challenges related to information fragmentation, lack of coordination, and inefficiencies in project delivery. In recent years, digital transformation initiatives have introduced new approaches to improve information management, with the Common Data Environment (CDE) emerging as a central tool. A CDE is a digital collaborative platform that facilitates the collection, management, and dissemination of all project-related information among stakeholders in a controlled and standardised environment (ISO 19650, 2018). Within Brazil's construction sector, the development and adoption of CDEs are closely tied to the implementation of Building Information Modelling (BIM) and the national BIM mandate under the Estratégia BIM BR (BIM BR Strategy).

The Brazilian construction sector is traditionally fragmented, with stakeholders ranging from architects and engineers to contractors and public agencies often working in silos. This fragmentation leads to communication breakdowns, design inconsistencies, and costly delays (Souza & Kagioglou, 2019). CDEs have been promoted as a solution to overcome these challenges by creating a single source of truth for project data. By consolidating information in a structured and accessible manner, CDEs improve transparency, reduce rework, and ensure that project participants are working from up-to-date and accurate information (Antunes & Scheer,

2020). In Brazil, the growing use of BIM has accelerated the recognition of CDEs as a necessary backbone for collaborative digital practices.

The government-led Estratégia BIM BR, introduced in 2018, provided an important push for the digitalisation of Brazil's construction industry. The strategy outlined phased targets for BIM adoption in public projects, starting in 2021 with design stages and extending to construction and asset management in later phases. This roadmap emphasised the need for standardised processes and collaborative environments, effectively laying the groundwork for CDE implementation (Costa & Furlani, 2020). For example, federal agencies managing infrastructure and building projects have increasingly required contractors to adopt BIM-enabled workflows supported by CDEs, thereby embedding collaborative digital platforms into procurement practices.

From an industry perspective, large Brazilian contractors and multinational firms have taken the lead in adopting CDEs. These companies have implemented platforms such as Autodesk BIM 360, Trimble Connect, and local solutions tailored to Brazil's regulatory and operational environment. Case studies have shown that firms using CDEs report improved project coordination, reduced design clashes, and better document management (Barison & Santos, 2011; Vazquez-Corral et al., 2020). However, the use of CDEs is not yet widespread among small and medium-sized enterprises (SMEs), which dominate Brazil's construction market. The primary barriers include high implementation costs, limited technical knowledge, and resistance to changing established workflows.

Costa and Furlani's (2021) research has highlighted that the adoption of CDEs in Brazil is still at a formative stage, with practices often inconsistent across projects. One critical challenge is the lack of standardised data management protocols that ensure interoperability between different software platforms (Antunes & Scheer, 2020). While the ISO 19650 standards provide international guidelines, their localisation to the Brazilian context remains a work in progress. Furthermore, regional disparities exist, with CDE use more common in metropolitan areas such as São Paulo and Rio de Janeiro compared to smaller cities, where digitalisation levels are lower (Souza & Kagioglou, 2019).

Despite these challenges, the potential benefits of CDEs in Brazil are considerable. Beyond improving coordination, CDEs enable better lifecycle management of assets by integrating design, construction, and facilities management data. This is particularly relevant in Brazil, where the government invests heavily in public infrastructure projects that require long-term asset maintenance. Additionally, the integration of CDEs with advanced technologies such as digital twins, cloud computing, and IoT devices could enhance predictive maintenance and operational efficiency in the Brazilian built environment (Vazquez-Corral et al., 2020).

Capacity-building initiatives are critical to supporting the expansion of CDE adoption. Brazilian universities and professional associations have increasingly introduced BIM and CDE-related modules in curricula and training programs, equipping future professionals with digital collaboration skills (Barison & Santos, 2011). Collaborative research projects between academia and industry also play a role in testing CDE applications in areas such as infrastructure delivery, sustainable design, and smart cities. Such efforts are vital for bridging the gap between policy intentions and practical implementation in Brazil's construction sector.

Consequently, while the Common Data Environment is still in its early stages of adoption in Brazil, it is becoming a cornerstone of the country's broader digital construction transformation. The alignment of government policy, industry innovation, and academic training provides a strong foundation for growth. To fully realise the benefits of CDEs, Brazil must address challenges related to cost, interoperability, and cultural resistance, while expanding support to SMEs. If these barriers are overcome, CDEs could significantly improve transparency, efficiency, and sustainability in Brazil's construction industry, positioning the sector to meet future demands more effectively.

5.18 State of the Construction Industry in Papua New Guinea

Papua New Guinea's (PNG) construction industry is shaped by the country's challenging geography, dispersed population, and a development trajectory tied closely to resource booms. On the demand side, the sector expands in cycles alongside large extractives and public infrastructure programs, especially roads that knit together mountainous and rainforest regions, but faces cost, governance, and environmental-planning headwinds that limit steady capacity growth (Alamgir et al., 2019). Urbanisation amplifies both need and complexity: Port Moresby and other urban centres have grown rapidly, with a large share of residents living in informal settlements that lack serviced land, utilities, and legal tenure. This produces a persistent backlog for basic infrastructure—roads, water, power, drainage—and affordable housing, all of which places sustained pressure on public works and building activity (Ezebilo & Savadogo, 2021; Jones, 2012).

Physical and environmental constraints are fundamental industry cost drivers. Road and bridge projects traverse steep, erosion-prone terrain and high-rainfall catchments, where access, mobilization, and maintenance are expensive, and construction footprints can open new deforestation frontiers if not carefully sited and sequenced (Alamgir et al., 2019). These conditions make logistics a major portion of project budgets, complicate scheduling, and heighten lifecycle risks, which in turn raise required margins and thin the pool of contractors with the capability and equipment to bid and deliver. In urban areas, the scarcity of serviced, titled land inflates project lead times and transaction costs, while the co-existence of customary and state land systems requires extended stakeholder processes before construction can commence (Ezebilo & Savadogo, 2021).

Labour-market structure is another binding constraint. Historically, PNG has relied on a small cohort of larger firms and many small indigenous contractors, with the latter often facing constraints in finance, management systems, and technical capability, leading to variability in cost, time, and quality performance (Wasi &

Skitmore, 2001). Skills gaps in trades and site supervision persist, reflecting uneven technical and vocational education and training (TVET) outcomes and limited pathways for structured on-the-job learning on complex projects. The result is periodic dependence on expatriate specialists during project spikes, with limited diffusion of skills across the wider industry between cycles (Wasi & Skitmore, 2001).

Urbanisation patterns strongly influence the project pipeline and delivery modalities. Research on Port Moresby's informal settlements shows residents prioritise core infrastructure—potable water, health facilities, access roads and electricity—and support upgrading when they are meaningfully engaged in planning and implementation. Sequencing road access first is often essential to unlock subsequent utility construction efficiently (Ezebilo & Savadogo, 2021). For the construction industry, these findings imply that participatory settlement upgrading, incremental regularisation, and serviced-land development can generate steady, multi-year programs well suited to building local contractor capability and stabilising demand beyond mega-projects (Ezebilo & Savadogo, 2021; Jones, 2012).

Governance and coordination challenges ripple through the sector. Jones (2012) documents fragmented urban decision-making and weak inter-agency coordination that slow approvals and dilute accountability, raising delivery risk premia. In infrastructure, the absence—at times—of integrated land-use, environmental, and transport planning fosters project-by-project decision-making that can raise long-run maintenance burdens or trigger environmental externalities, especially in frontier road building (Alamgir et al., 2019). Strengthening upstream planning, adopting context-appropriate standards, and investing in maintenance regimes would bolster whole-of-life value and create a more predictable workflow for contractors.

Looking ahead, PNG's construction outlook is best understood as a series of overlapping opportunities conditioned by capability and governance reforms. Urban service backlogs, climate-resilient transport upgrades, school and health facility programs, and incremental housing delivery can provide a diversified, demand-stable base. Realising this potential will depend on: (i) structured capacity-building for small and medium local contractors (financial management, safety, QA/QC, and site supervision); (ii) demand-led TVET tied to project pipelines; (iii) stronger urban governance to expedite serviced-land and permitting; and (iv) environmentally informed route selection and maintenance planning to reduce lifecycle costs and unintended deforestation impacts (Alamgir et al., 2019; Ezebilo & Savadogo, 2021; Jones, 2012; Wasi & Skitmore, 2001). With these shifts, PNG's construction industry can move from boom-bust cyclicality toward a more resilient, inclusive growth path that better matches local skills formation to long-term national infrastructure needs.

5.19 Construction Information Management in Papua New Guinea

Construction information management (CIM) in Papua New Guinea (PNG) is still at an early stage of development compared with more industrialised economies. The country's construction sector is characterised by fragmented supply chains, diverse project delivery methods, and significant reliance on small and medium-sized indigenous contractors, all of which present challenges for information flow, coordination, and documentation. Historically, project communication and documentation in PNG have been dominated by paper-based systems, contractual correspondence, and traditional site records, with limited adoption of digital platforms for real-time data sharing (Wasi & Skitmore, 2001). This results in inefficiencies such as delays in approvals, cost variations, and difficulties in monitoring project progress and quality compliance.

One of the primary barriers to effective CIM in PNG lies in institutional capacity. Many small indigenous contractors, who make up a large share of the industry, lack structured management systems and trained personnel to handle complex project documentation, cost control, and reporting frameworks (Wasi & Skitmore, 2001). Without robust information management practices, these firms often face challenges in project scheduling, cash-flow management, and contractual claims, limiting their competitiveness and growth potential. At the same time, larger international contractors engaged in major resource and infrastructure projects often deploy more advanced management systems, but the transfer of these practices to the local ecosystem remains minimal due to limited integration and skills diffusion (Jones, 2012).

Urbanisation and the rapid expansion of informal settlements in Port Moresby and other urban centres create additional information management complexities. For example, infrastructure upgrading projects in settlements require significant stakeholder engagement, community mapping, and iterative planning—processes that demand robust data collection, storage, and sharing mechanisms (Ezebilo & Savadogo, 2021). In many cases, however, information is not systematically captured, making it difficult to track service provision, monitor implementation, or coordinate agencies. Weak inter-agency data integration has been highlighted as a critical governance bottleneck, as urban authorities often operate in silos with limited information sharing, leading to duplication and delays in project delivery (Jones, 2012).

Geographic and environmental conditions also influence information management needs. Infrastructure projects that traverse PNG's rugged landscapes face logistical risks such as landslides, flooding, and access challenges (Alamgir et al., 2019). Effective construction information systems could support risk assessments, environmental monitoring, and maintenance planning. However, limited investment in digital platforms, geographic information systems (GIS), and data-driven project management tools means that decisions are often reactive rather than evidence-based. This shortfall raises lifecycle costs and constrains the industry's ability to adapt to climate and sustainability imperatives.

Looking forward, improving CIM in PNG requires targeted strategies. First, capacity-building initiatives for small and medium contractors should include training in digital project management tools, document control, and contract administration. Second, government agencies can play a central role by mandating the use of integrated information systems in public procurement, gradually encouraging standardisation and transparency. Third, leveraging community-based data collection and participatory mapping in urban upgrading projects can enhance inclusivity while generating useful datasets for planners and contractors. Finally, partnerships with universities and regional training institutions could accelerate knowledge transfer in construction informatics, GIS applications, and building information modelling (BIM), laying the groundwork for gradual digital transformation.

Construction information management in PNG is constrained by fragmented practices, limited technological adoption, and capacity gaps. While large contractors have introduced structured systems on major projects, the broader sector—dominated by small indigenous firms—struggles with basic project documentation and communication. With urbanisation pressures, environmental risks, and the need for efficiency, advancing CIM is essential for the sector's modernisation. By investing in training, standardisation, and digital platforms, PNG can strengthen its construction sector and align it with international best practices while fostering local capability development.

5.20 Building Information Modelling in Papua New Guinea

Building Information Modelling (BIM) in Papua New Guinea (PNG) is still in its infancy compared to global trends, with very limited adoption across the construction industry. Unlike developed countries where BIM has become a standard for digital design, collaboration, and lifecycle asset management, PNG's construction sector is largely dependent on traditional methods of project delivery, documentation, and communication (Wasi & Skitmore, 2001). Most projects continue to rely on two-dimensional (2D) drawings, paper-based records, and conventional contract management processes, which limit efficiency and hinder collaboration. The dominance of small and medium-sized indigenous contractors, who often lack the technological and financial capacity to invest in advanced digital tools, contributes significantly to the slow uptake of BIM in the country (Jones, 2012).

Despite this, the potential for BIM in PNG is significant, particularly as the country faces urbanisation pressures, infrastructure deficits, and complex environmental conditions. BIM has the capacity to address some of PNG's recurring challenges, such as poor coordination among stakeholders, cost overruns, and project delays. For example, 3D and 4D BIM models could provide more accurate planning for infrastructure projects across PNG's challenging terrain, enabling better simulation of risks such as landslides and flooding (Alamgir et al., 2019). Moreover, the use of BIM could strengthen asset management and facilities maintenance in urban settlements, where fragmented service delivery and poor data integration currently undermine sustainable development (Ezebilo & Savadogo, 2021).

Institutional and policy frameworks, however, remain a key barrier. Unlike countries such as Singapore or the UK, where government mandates have accelerated BIM adoption, PNG lacks national construction standards, digital transformation strategies, or regulatory incentives to encourage BIM integration. Public procurement processes are still paper-driven, and there is little evidence of coordinated digital infrastructure policies in the construction sector (Jones, 2012). As a result, BIM adoption, where it occurs, is limited to international contractors or donor-funded megaprojects that bring in global expertise and technology but do not significantly diffuse knowledge to the wider local industry.

Capacity development is another major issue. There is a shortage of professionals trained in BIM software such as Revit, Navisworks, or ArchiCAD in PNG. The country's tertiary education and vocational training institutions have not yet integrated BIM into construction-related curricula, which limits exposure for future architects, engineers, and quantity surveyors. Without structured training and local expertise, reliance on foreign consultants persists, creating dependency and limiting the long-term sustainability of BIM use (Wasi & Skitmore, 2001).

Nevertheless, opportunities exist for gradual BIM introduction in PNG. International development projects funded by multilateral agencies such as the Asian Development Bank or the World Bank could serve as catalysts by embedding BIM standards in infrastructure projects. Universities and technical training institutions could establish partnerships with regional neighbours, such as Australia and New Zealand, to build local BIM capability. Additionally, cloud-based BIM platforms offer opportunities for PNG to leapfrog traditional IT infrastructure investments by enabling collaborative work across dispersed geographies. This could be particularly useful in overcoming PNG's unique logistical and geographic barriers to construction delivery.

BIM adoption in PNG remains limited, hindered by institutional gaps, lack of skills, and minimal policy support. However, as infrastructure demand grows and urbanisation intensifies, BIM provides a viable pathway for improving efficiency, collaboration, and long-term asset management in the construction sector. By leveraging international partnerships, embedding BIM training in education, and integrating BIM into government procurement frameworks, PNG could gradually modernise its construction industry and align it with global digital practices.

5.21 Common Data Environment in Papua New Guinea's Construction Industry

The concept of a Common Data Environment (CDE) in Papua New Guinea's (PNG) construction industry remains largely underdeveloped. A CDE refers to a shared digital platform where construction project information—including drawings, models, documents, and schedules—can be collected, managed, and disseminated among stakeholders in a structured manner (ISO 19650, 2018). In developed countries, such as the United Kingdom or Australia, CDEs have become integral to project delivery, especially with the integration of Building Information Modelling (BIM). However, in PNG, the adoption of both BIM and supporting platforms like CDEs is still at an early stage, with the industry largely reliant on traditional paper-based and fragmented information management systems (Wasi & Skitmore, 2001).

One of the main barriers to CDE adoption in PNG is the dominance of small and medium-sized indigenous contractors who often lack the financial and technological capacity to invest in advanced digital systems (Wasi & Skitmore, 2001). Information exchange on projects is still carried out via physical documentation, emails, and phone calls, leading to inefficiencies such as duplicated work, delays, and miscommunication. The absence of national digital standards for construction information management exacerbates the problem, as there are no regulatory frameworks or government-led initiatives to mandate the use of CDEs in public or private sector projects (Jones, 2012).

Despite these challenges, the benefits of CDE adoption in PNG's construction sector are considerable. A CDE could provide a central repository of project data accessible to contractors, consultants, clients, and government agencies, thereby reducing the fragmentation that currently hampers project delivery. This is especially relevant in PNG, where geographic and logistical barriers make physical coordination among project stakeholders difficult. A CDE could facilitate real-time collaboration, streamline approval processes, and improve transparency in procurement and contract administration, which are areas that have historically faced issues of inefficiency and corruption (Ezebilo & Savadogo, 2021).

Moreover, donor-funded infrastructure projects by international agencies such as the Asian Development Bank and the World Bank represent opportunities for introducing CDE practices. These projects often involve international contractors and consultants who are familiar with BIM-enabled workflows and CDE platforms. While such expertise is sometimes imported without widespread diffusion to local firms, there is potential for knowledge transfer through collaborative project structures. Embedding CDE use as a contractual requirement in these projects could create gradual exposure and capacity development among local stakeholders (Alamgir et al., 2019).

Another key issue is the lack of skilled professionals trained in digital construction tools. PNG's universities and vocational institutions have yet to integrate BIM and digital information management into their curricula, limiting awareness of CDEs among graduates in architecture, engineering, and construction management. Without adequate training, reliance on traditional methods will persist. Building local capacity in digital construction through academic programs, short courses, and partnerships with regional neighbours like Australia and New Zealand could play a critical role in introducing CDEs to the industry.

Cloud-based CDE platforms also present an opportunity for PNG to leapfrog traditional infrastructure constraints. By using cloud solutions, project teams dispersed across PNG's challenging terrain could access the same up-to-date project data, eliminating the need for physical transfer of documents. This could improve project delivery efficiency and reduce costly errors due to outdated or inconsistent information. However, such solutions would require addressing PNG's limited ICT infrastructure, particularly unreliable internet connectivity in rural areas, which currently poses a barrier to digital collaboration (Jones, 2012).

However, while the adoption of a Common Data Environment in PNG's construction industry is currently minimal, its potential benefits for improving project coordination, transparency, and efficiency are significant. The path forward lies in creating enabling conditions through policy frameworks, integrating digital construction education, leveraging donor-funded projects for technology transfer, and investing in ICT infrastructure. If these barriers are systematically addressed, CDEs could transform PNG's construction industry by aligning it with global best practices in digital project delivery.

5.22 Summary

This chapter has examined construction information management in emerging economies, with a focus on South Africa, India, Egypt, Brazil, and Papua New Guinea. Each country's construction industry was first contextualised by outlining its current state, challenges, and growth potential. Across these nations, construction remains a significant driver of economic development, yet it is consistently hindered by fragmented processes, cost and time overruns, and weak regulatory frameworks. The analysis of construction information management revealed that while traditional methods still dominate, there is a gradual shift toward digital systems. In South Africa and India, information management is evolving through government initiatives and private sector investments, whereas in Egypt and Brazil adoption remains slower due to structural and institutional barriers. Papua New Guinea lags behind, with information still largely managed through manual and paper-based processes. Building Information Modelling (BIM) emerged as a key transformative tool across these countries. South Africa and India show relatively stronger adoption, supported by policy frameworks and pilot projects. Egypt and Brazil demonstrate growing but inconsistent uptake, while in Papua New Guinea BIM adoption is minimal and heavily reliant on foreign contractors. The discussion on the Common Data Environment (CDE) highlighted its potential to enhance collaboration, data sharing, and project delivery efficiency. However, adoption remains uneven, with South Africa and India leading, Brazil and Egypt in transitional phases, and Papua New Guinea at a very early stage. The chapter consequently illustrates both the opportunities and constraints in digitising construction information management in emerging economies, underscoring the need for regulatory support, capacity building, and infrastructure investment to enable widespread transformation.

References

Adekunle, P., Aigbavboa, C., Akinradewo, O., Oke, A., & Aghimien, D. (2022a). Construction Information Management: Benefits to the Construction Industry. *Sustainability, 14*(18), 11366. https://doi.org/10.3390/su141811366

Adekunle P, Aigabvboa C, Thwala W, Akinradewo O and Oke A (2022b) Challenges confronting construction information management. *Frontiers in Built Environment, 8*, 1075674. https://doi.org/10.3389/fbuil.2022.1075674

Adekunle P, Aigbavboa C, Thwala D, Oke A, Akinradewo O (2022c) Construction information management: the role of fourth industrial revolution tools. *Human Factors in Architect Sustain Urban Plann Infrastruct* 254–261. https://doi.org/10.54941/ahfe1002359

Adekunle, P., Aigbavboa, C., Oke, A., Akinradewo, O., & Otasowie, O. (2023). Application of big data and internet of things in the built environment: a bibliometric review. Proceedings of the 31st Annual Conference of the International Group for Lean Construction (IGLC31), 50–58. https://doi.org/10.24928/2023/0138

Adekunle, P., Aigbavboa, C., Otasowie, K., & Ikuabe, M. (2023). Evaluating the Implementation of Machine Learning Techniques in the South African Built Environment. In S. M. Ahmed, S. Azhar, A. D. Saul, K. L. Mahaffy, & R. U. Farooqui (Eds.), Proceedings of the 13th International Conference on Construction in the 21st Century, CITC 2023 (International Conference on Construction in the 21st Century; Vol. 2023-May). East Carolina University.

Adekunle P, Aigbavboa C, Akinradewo O, Ikuabe M, Otasowie K. (2024a) Towards the Uptake of Digital Technologies for Construction Information Management: A Partial Least Squares Structural Equation Modelling Approach. *Buildings; 14*(3):827. https://doi.org/10.3390/buildings14030827

Adekunle, P., Aigbavboa, C., Akinradewo, O., Otasowie, K., Adekunle, S. (2024b). Methods of Managing Construction Information in the Fourth Industrial Revolution Era. In: Papadikis, K., Zhang, C., Tang, S., Liu, E., Di Sarno, L. (eds) Towards a Carbon Neutral Future. ICSBS 2023. Lecture Notes in Civil Engineering, vol 393. Springer, Singapore. https://doi.org/10.1007/978-981-99-7965-3_41

Adekunle, S., Adekunle, P., Aigbavboa, C., Akinradewo, O., & Ebekozien, A. (2024c). Benefits of Emerging Technologies Adoption for Health and Safety (HSE) by Small and Medium-Sized Enterprises (SMEs). International Conference on Construction in the 21st Century.

Adekunle, P., Aigbavboa, C., Otasowie, K., Ikuabe, M. (2024d). Leveraging on Technologies Within Quality Management Systems for Enhanced Project Success in Construction. In: Arai, K. (eds) Proceedings of the Future Technologies Conference (FTC) 2024, Volume 4. FTC 2024. Lecture Notes in Networks and Systems, vol 1157. Springer, Cham. https://doi.org/10.1007/978-3-031-73128-0_32

Adekunle, P., Aigbavboa, C., Ikuabe, M., Otasowie, K. (2024e). Challenges Hindering the Promotion of Machine-Learning Techniques in the Construction Industry. In: Rotimi, J.O.B., Shahzad, W.M., Sutrisna, M., Kahandawa, R. (eds) Advances in Engineering Project, Production, and Technology. EPPM 2023. Lecture Notes in Mechanical Engineering. Springer, Cham. https://doi.org/10.1007/978-3-031-56878-7_21

Aigbavboa, C., Adekunle, P., Ikuabe, M., Otasowie, K. (2024). Digitisation of the Construction Industry for Economic Growth: Implementation Strategies. In: Arai, K. (eds) Proceedings of the Future Technologies Conference (FTC) 2024, Volume 4. FTC 2024. Lecture Notes in Networks and Systems, vol 1157. Springer, Cham. https://doi.org/10.1007/978-3-031-73128-0_34

Akinradewo, O., Hafez, M., Aigbavboa, C. Ebekozien, A. Adekunle, P. Otasowie, O. (2024) Innovating Built Environment Education to Achieve SDG 4: Key Drivers for Integrating Augmented Reality Technologies. Sustainability, 16, 8315. https://doi.org/10.3390/su16198315

Akintola, A., Root, D., & Venkatachalam, S. (2017). Key constrainsts to optimal and widespread implementation of BIM in the South African construction industry. In *Thirty-Third Annual Conference* (Vol. 25, p. 75).

Amaral, T. G., Ghisi, E., & Mendes, N. (2019). Sustainability assessment of buildings in Brazil: Trends, challenges, and opportunities. *Journal of Cleaner Production, 234*, 1234–1247.

Antunes, R., & Scheer, S. (2020). BIM implementation in the Brazilian public sector: Advances and challenges. *Journal of Information Technology in Construction (ITcon), 25*, 232–247.

Arditi, D., Fischman, G., & Sivaloganathan, S. (2000). Performance-based management of construction projects. *International Journal of Project Management, 18*(4), 185–195. https://doi.org/10.1016/S0263-7863(99)00027-3

Armoed, Z., & Naidu, D. E. (2024). The rate of building information modeling adoption at higher education institutions in South Africa. *Journal of Construction Research, 16*(02), 35–45.
Arora, A., & Awasthi, A. (2020). Adoption of BIM and digital technologies in the Indian construction industry. *International Journal of Construction Management, 20*(8), 783–794.
Attia, S., & Ibrahim, A. (2018). Towards a sustainable building sector in Egypt: A roadmap for policy and practice. *Journal of Cleaner Production, 183*, 83–95.
Azis, R., Hafez, S., & El-Rayes, K. (2004). Knowledge-based scheduling systems in construction projects. *Journal of Construction Engineering and Management, 130*(4), 546–554. https://doi.org/10.1061/(ASCE)0733-9364(2004)130:4(546)
Azhar, S. (2011). Building information modeling (BIM): trends, benefits, risks, and challenges for the AEC industry. *Leadership Management in Engineering, 11*(3), 241–252. https://doi.org/10.1061/(ASCE)LM.1943-5630.0000127
Barison, M. B., & Santos, E. T. (2011). BIM teaching strategies: An overview of the current approaches. In *Proceedings of CIB W78-W102 Conference*, Sophia Antipolis, France.
Bisschoff, H., & Benade, S. J. (2008). *Partnering on information management projects in South Africa* (pp. 1363–1370). IEEE.
Bryceson, D. F., & Mackinnon, D. (2012). Urbanization, infrastructure and construction: Global trends and challenges. *International Journal of Construction Management, 12*(2), 112–122. https://doi.org/10.1080/15623599.2012.673466
Chimhundu, S. (2016). *A study on the BIM adoption readiness and possible mandatory initiatives for successful implementation in South Africa.* Master's thesis, University of the Witwatersrand, Johannesburg.
Construction Industry Board of Denmark (CIBD). (2004). Construction industry report 2004: Information management and industry performance. CIBD.
Construction Industry Board of Denmark (CIBD). (2012). Digital transformation in the Danish construction industry: Industry status report 2012. CIBD.
Costa, D. B., & Furlani, T. (2020). BIM implementation in Brazil: Challenges and opportunities. *Journal of Information Technology in Construction (ITcon), 25*, 218–247.
Desouza, K. (2009). Securing information assets: The great information game. *Business Information Review, 26*(1), 35–41.
Dolla, T., Venkatachalam, S., & Kumar, V. S. (2024). Institutional shaping of CDE implementation in BIM-enabled AEC projects. *Journal of Information Technology in Construction (ITcon), 29*, 826–849.
Doloi, H., Sawhney, A., Iyer, K. C., & Rentala, S. (2012). Analysing factors affecting delays in Indian construction projects. *International Journal of Project Management, 30*(4), 479–489.
El-adaway, I. H., Abotaleb, I. S., & Kandil, A. (2020). Construction industry in Egypt: Status, challenges, and the future outlook. *Journal of Management in Engineering, 36*(3), 05020002.
Froise, T., & Shakantu, W. (2014). Diffusion of innovations: An assessment of building information modelling uptake trends in South Africa. *Journal of Construction Project Management and Innovation, 4*(2), 895–911.
Ghosh, S. (2021). Impact of COVID-19 on migrant workers and the construction industry in India. *Economic and Political Weekly, 56*(20), 34–40.
Gu, N. & London, K.. (2010). Understanding and facilitating BIM adoption in the AEC industry. *Automation in Construction, 19*(8), 988–999. https://doi.org/10.1016/j.autcon.2010.06.009
Gupta, R. (2020). Sustainable construction practices in India: Trends and challenges. *Journal of Cleaner Production, 258*, 120–146.
Harris, F. & Nawawi, A. H.. (2013). ICT adoption in construction project management: Malaysian contractors' perception. *Journal of Engineering, Project, and Production Management, 3*(1), 1–12. https://doi.org/10.32738/JEPPM.201301.0001
Haron, A.T., Marshall-Ponting A.J., Aouad G.F, 2010. *Building information modelling: Literature review on model to determine the level of uptake by the organisation,* Manchester, United Kingdom (pp. 1–19).

Hegazy, T., & El-adaway, I. (2017). Challenges facing the Egyptian construction industry: An analytical perspective. *International Journal of Construction Management, 17*(4), 326–338.
Kaminski, J. (2011). Diffusion of innovation theory. *Canadian Journal of Nursing Informatics, 6*(2), pp. 1–7. https://cjn.net/journal/p=144
Ikuabe, M., Aigbavboa, C., Ebekozien, A., Otasowie, K., Adekunle, P. (2025a). Navigating the Hurdles of Implementing Remote Construction Management in Developing Economies. In: GhaffarianHoseini, A., Ghaffarianhoseini, A., Rahimian, F., Babu Purushothaman, M. (eds) Proceedings of the International Conference on Smart and Sustainable Built Environment (SASBE 2024). SASBE 2024. Lecture Notes in Civil Engineering, vol 591. Springer, Singapore. https://doi.org/10.1007/978-981-96-4051-5_159
Ikuabe, M., Aigbavboa, C., Adekunle, P., Otasowie, K. (2025b). Digital Transformation in the Construction Industry: Mapping Out the Required Leadership Attributes. In: GhaffarianHoseini, A., Ghaffarianhoseini, A., Rahimian, F.,Babu Purushothaman, M. (eds) Proceedings of the International Conference on Smart and Sustainable Built Environment (SASBE 2024). SASBE 2024. Lecture Notes in Civil Engineering, vol 591. Springer, Singapore. https://doi.org/10.1007/978-981-96-4051-5_157
ISO 19650. (2018). Organization and digitization of information about buildings and civil engineering works, including building information modelling (BIM) — Information management using building information modelling. ISO.
Kekana, T. G., Aigbavboa, C. O., & Thwala, W. D. (2014). *Building information modelling (BIM): Barriers in adoption and implementation strategies in the South Africa construction industry,* Pattaya, Thailand (pp. 1–3).
Kubjana, L., Adekunle, P., Aigbavboa, C., Otasowie, K. (2024). Digitisation in the South African Construction Industry: Recurring Challenges. In: Arai, K. (eds) Proceedings of the Future Technologies Conference (FTC) 2024, Vol 4. FTC 2024. Lecture Notes in Networks and Systems, vol 1157. Springer, Cham. https://doi.org/10.1007/978-3-031-73128-0_35
Kumar, R., & Bansal, V. (2020). The role of the construction industry in India's economic development. *Journal of Infrastructure Development, 12*(1), 1–17.
Lima, F., & da Silva, C. (2019). Corruption in the Brazilian construction industry: Lessons from Operation Car Wash. *Construction Management and Economics, 37*(10), 545–561.
Maconachie, L. (2012). Information systems and efficiency in the construction sector. *Construction Management and Economics, 30*(7), 485–494. https://doi.org/10.1080/01446193.2012.678432
Maconachie, L.. (2012). Corporate information systems and construction industry efficiency. *Construction Management and Economics, 30*(7), 485–494. https://doi.org/10.1080/01446193.2012.678432
Maritz, T., Klopper, C., & Siglé, T. (2005). Developing a national standard/code of practice for the classification of construction information in South Africa. *Building and Environment, 40*(7), 1003–1009.
Marzouk, M., & El Kherbawy, A. (2019). Megaprojects in Egypt: Planning, financing, and challenges. *Built Environment Project and Asset Management, 9*(5), 639–654.
Mattos, C. (2016). Construction industry performance in Brazil: An overview. *Built Environment Project and Asset Management, 6*(5), 478–493.
Mishra, A., & Jha, K. N. (2021). Public-private partnerships in Indian infrastructure development: Performance and prospects. *Built Environment Project and Asset Management, 11*(4), 623–639.
Mpingana, W. B., Akinradewo, O., & Aigbavboa, C. (2025). Exploring critical success factors for cloud data management in the South African construction industry. In *Advances in Information Technology in Civil and Building Engineering—Proceedings of ICCCBE 2024.* Lecture Notes in Civil Engineering (Vol. 629, pp. 652–667). Springer.
Oliveira, R., Lima, J., & Andrade, A. (2021). Occupational safety in Brazilian construction: An analysis of risks and accidents. *Safety Science, 139,* Article 105254.
Osman, H., Fayek, A. R., & Marzouk, M. (2021). Public-private partnerships in Egypt: Opportunities and challenges for infrastructure development. *Journal of Construction Engineering and Management, 147*(6), 04021047.

Otasowie, K., Aigbavboa, C., Oke, A., Adekunle, P., Zanele, B. (2023a). Blockchain Technology Adoption in Construction Projects Logistics in South Africa: Militating Factors. In: Arai, K. (eds) Proceedings of the Future Technologies Conference (FTC) 2023, Volume 3. FTC 2023. Lecture Notes in Networks and Systems, vol 815. Springer, Cham. https://doi.org/10.1007/978-3-031-47457-6_21

Otasowie, K., Aigbavboa, C., Ikuabe, M., Adekunle, P., Oke A., & Adekunle, S. (2023b). Barriers to the adoption of building information modeling in quantity surveying practice in South Africa. Proceedings of the 31st Annual Conference of the International Group for Lean Construction (IGLC31), 41–49. https://doi.org/10.24928/2023/0137

Otasowie, K., Aigbavboa, C., Ikuabe, M., Adekunle, P. (2024). Big Data Adoption in Construction and Demolition Waste Management: Prevailing Challenges in Developing Nations. In: Arai, K. (eds) Proceedings of the Future Technologies Conference (FTC) 2024, Volume 4. FTC 2024. Lecture Notes in Networks and Systems, vol 1157. Springer, Cham. https://doi.org/10.1007/978-3-031-73128-0_42

Otasowie, K., Aigbavboa, C., Oke, A., Adekunle, P. (2024). Influencing Factors for Blockchain Adoption in Construction Project Logistics. In: Li, D., Zou, P.X.W., Yuan, J., Wang, Q., Peng, Y. (eds) Proceedings of the 28th International Symposium on Advancement of Construction Management and Real Estate. CRIOCM 2023. Lecture Notes in Operations Research. Springer, Singapore. https://doi.org/10.1007/978-981-97-1949-5_130

Peci, A., Cavalcanti, B., & Lotta, G. (2020). Public-private partnerships in Brazil: Advances, challenges, and prospects. *Public Money & Management, 40*(6), 446–454.

Sacks, R., Eastman, C., Lee, G., & Teicholz, P. (2018). *BIM handbook: A guide to building information modeling for owners, managers, designers, engineers and contractors* (3rd ed.). Wiley.

Sanches, R., & Vieira, L. (2018). The role of construction in Brazil's economic growth: An empirical analysis. *Economic Analysis and Policy, 60*, 45–56.

Sharma, S., & Sharma, R. (2019). Megaprojects and urban transformation in India. *Cities, 94*, 121–129.

Silva, F., Rangel, R., & Moura, P. (2018). Housing policies and social inclusion in Brazil: Assessing the impact of "Minha Casa, Minha Vida." *Habitat International, 73*, 1–9.

Soliman, A., El Gohary, N., & Kandil, A. (2018). BIM adoption in the Egyptian construction industry: Barriers and drivers. *International Journal of Construction Education and Research, 14*(4), 250–266.

Souza, R., & Kagioglou, M. (2019). Productivity in the Brazilian construction industry: Challenges and improvement opportunities. *International Journal of Productivity and Performance Management, 68*(4), 731–747.

Succar, B. (2009). Building information modelling framework: A research and delivery foundation for industry stakeholders. *Automation in Construction, 18*(3), 357–375.

Tambwe, O. T., Aigbavboa, C. O., Akinradewo, O. I., & Adekunle, P. A. (2025). Measures to Address Cyber-Attacks in Construction Project Data Management Processes: A Cybersecurity Perspective. IET Information Security, 2025(1). https://doi.org/10.1049/ise2/7398742

The Express Wire. (2020, June 5). Digital construction trends in 2020: What contractors need to know. The Express Wire. https://www.theexpresswire.com/2020/06/05/digital-construction-trends-in-2020

Vazquez-Corral, J., de Andrade, M., & Cardoso, D. (2020). BIM-GIS integration for infrastructure projects in Brazil: Current practices and future opportunities. *Automation in Construction, 113*, Article 103148.

Windapo, A. O., & Cattel, K. (2013). The South African construction industry: Perceptions of key challenges facing its performance, development and growth. *Journal of Construction in Developing Countries, 18*(2), 65.

Zheng, L., & London, K. (2010). Digital technologies in managing construction information flows. *Automation in Construction, 19*(8), 970–979. https://doi.org/10.1016/j.autcon.2010.06.008

Chapter 6
Conclusion

Abstract This chapter summarises the study's findings showing how the research objectives have been achieved and the research questions answered. The chapter further discusses the research's contribution to knowledge made based on the study's findings and the areas for further research.

6.1 Introduction

This chapter summarises the study's findings showing how the research objectives have been achieved and the research questions answered. The chapter further discusses the research's contribution to knowledge made based on the study's findings and the areas for further research.

6.2 The Research's Contribution of Knowledge

This study revealed the need for infrastructure to support the development of an effective and efficient construction information management system. This would enable educational construction enterprises to be more imaginative and create better opportunities for increasing information flow and providing timely information to clients and other end users. It was also demonstrated that construction information management allows for the creation of a conducive environment for ensuring trustworthy results and enhancing top-level management's decision-making process. As a result, it is suggested that the use of cutting-edge technology in construction information management be increased to make it a standard practice in the sector. Policy should be implemented by regulatory and professional authorities to support the use of 4IR tools in the sector. There are still some in the construction sector that prefer the old management approach. Many activities in the industry have been slowed

© The Author(s), under exclusive license to Springer Nature Switzerland AG 2026
P. Adekunle et al., *Emerging Trends in Construction Information Management*,
https://doi.org/10.1007/978-3-032-09679-1_6

as a result of this. Finally, the study found that construction information management courses should be required in higher education programs to educate prospective construction professionals with a thorough understanding of construction data. Students must comprehend what information management entails, particularly in regard to the fourth industrial revolution, in order to quickly overcome construction information management issues.

6.3 Novelty of the Research

Construction information management improves project management, individual effort coordination, time management, and cost reduction. Misinformation in the construction business has frequently resulted in project delays, disagreements among stakeholders, and other unfavorable outcomes. Understanding construction information management will aid in standardizing the creation, storage, and management of information in order to facilitate collaboration between the contractor and client's representatives (architect, quantity surveyor, builder, civil engineer, project manager, mechanical and electrical engineer) and ensure that the client's information needs are met.

6.4 Limitation of the Study

During the implementation of this research investigation, a few drawbacks were discovered. To begin, the researcher was able to gather data, primarily from South African construction professionals. South Africa is a huge country with nine provinces, so visiting all of them was challenging. Furthermore, due to covid-19 limits, time constraints, and budgetary constraints, most provinces were unable to be visited in order to acquire data and figures that would further corroborate the findings of this study. As a result, the conclusions of this research study cannot be applied to a larger population. To do this, the study would need to be conducted on a larger, more representative sample of people from other developing nations. A larger, more diverse sample may also help to mitigate any potential bias introduced by a self-administered questionnaire.

6.5 Agenda for Future Research

The authors believe that a larger sample size would allow for even more gains. This will provide a more comprehensive understanding of how construction information management may be scaled up to meet the needs of a typical developing country construction industry. Future research can guarantee that a broader sample is chosen, one that spans more countries and includes more construction sector professionals. This will give you a better understanding of how information is managed in building projects. This will also enable results to be extrapolated to a larger population. Future research may look at the perceptions of potential clients (the developing countries population) about the methods that can help them communicate effectively with construction experts. This might then be passed on to the country's construction authorities.

6.6 Summary

The study's main goal was to evaluate construction data management in the South African construction industry. This was accomplished by reviewing construction information management approaches, identifying problems, weighing benefits, and establishing ways to improve construction information management. The study's initial conclusion was that the construction sector is always evolving as a result of new trends, increasing project-based work, technology, and enhanced construction processes, necessitating the implementation of a well-structured information management system. Due to the vast volume of recorded information during building operations, it was also concluded that the old information system is no longer suitable for modern construction processes. The current status of the construction business necessitates the development of systems for properly and quickly recording, processing, transferring, and storing data. The main advantages of construction information management are that it ensures statistics and figures to identify counterproductive procedures that cause project delays and procedures that improve productivity and efficiency for improved decision making, a drastic reduction in communication gaps, and a better understanding of client requirements. Finally, a good construction information management system has the potential to boost the value and long-term viability of the assets and facilities developed by the construction industry, allowing the republic's built environment to generate more societal value.

Bibliography

Abrams, L. S. (2010). Sampling hard to reach population in qualitative research populations in qualitative research: The case of incarcerated youth. *Qualitative Social Work, 9*(4), 536–550.

Adekunle, P., Aigbavboa, C., Akinradewo, O., Oke, A., & Aghimien, D. (2022). Construction information management: Benefits to the construction industry. *Sustainability, 14*, 11366. https://doi.org/10.3390/su141811366

Adekunle, P., Aigbavboa, C., Oke, A., Akinradewo, O., & Otasowie, O. (2023). Application of big data and internet of things in the built environment: A bibliometric review. In *Proceedings of the 31st Annual Conference of the International Group for Lean Construction (IGLC31)* (pp. 50–58). https://doi.org/10.24928/2023/0138

Adekunle, P., Aigbavboa, C., Otasowie, K., & Adekunle, S. (2023). Benefits of robotic utilization in the prefabricated construction industry. In *Proceedings of the 31st Annual Conference of the*

International Group for Lean Construction (IGLC31) (pp. 746–754). https://doi.org/10.24928/2023/0134

Adekunle, P., Aigbavboa, C., Otasowie, K., & Ikuabe, M. (2023). Evaluating the implementation of machine learning techniques in the South African built environment. In S. M. Ahmed, S. Azhar, A. D. Saul, K. L. Mahaffy, & R. U. Farooqui (Eds.), *Proceedings of the 13th International Conference on Construction in the 21st Century, CITC 2023* (International Conference on Construction in the 21st Century; Vol. 2023-May). East Carolina University.

Adekunle, P., Aigbavboa, C., Akinradewo, O., Ikuabe, M., & Otasowie, K. (2024a). Towards the uptake of digital technologies for construction information management: A partial least squares structural equation modelling approach. *Buildings, 14*(3), 827. https://doi.org/10.3390/buildings14030827

Adekunle, P., Aigbavboa, C., Akinradewo, O., Otasowie, K., & Adekunle, S. (2024b). Methods of managing construction information in the fourth industrial revolution era. In K. Papadikis, C. Zhang, S. Tang, E. Liu, & L. Di Sarno (Eds.), *Towards a Carbon Neutral Future. ICSBS 2023.* Lecture Notes in Civil Engineering (Vol. 393). Springer. https://doi.org/10.1007/978-981-99-7965-3_41

Adekunle, S., Adekunle, P., Aigbavboa, C., Akinradewo, O., & Ebekozien, A. (2024c). Benefits of emerging technologies adoption for health and safety (HSE) by small and medium-sized enterprises (SMEs). In *International Conference on Construction in the 21st Century.*

Aibinu, A. A., & Al-Lawati, A. M. (2010). Using PLS-SEM technique to model construction organisation willingness participate in e-bidding. *Automation in Construction, 19*, 714–724.

Aigbavboa, C., Adekunle, P., Ikuabe, M., & Otasowie, K. (2024). Digitisation of the construction industry for economic growth: Implementation strategies. In K. Arai (Eds.), *Proceedings of the Future Technologies Conference (FTC) 2024, Volume 4. FTC 2024.* Lecture Notes in Networks and Systems (Vol. 1157). Springer. https://doi.org/10.1007/978-3-031-73128-0_34

Ajoudani, A., Zanchettin, A. M., Ivaldi, S., Albu-Schäffer, A., Kosuge, K., & Khatib, O. (2018). Progress and prospects of the human-robot collaboration. *Automation in Robotics, 42*, 957–975. https://doi.org/10.1007/s10514-017-9677-2

Altieri, K. E., & Keen, S. L. (2019). Public health benefits of reducing exposure to ambient fine particulate matter in South Africa. *Science of the Total Environment, 684*, 610–620. https://doi.org/10.1016/j.scitotenv.2019.05.355

Andrew, F., & Les, R. (2007). *Defining the scope of construction sector* (pp. 1–45). The University of Salford UK.

Anon. (2011). Product development team stability and new product advantage: The role of decision-making process. *Journal of Marketing, 75*(1), 96–108.

Anon. (2020). *Information management.* [Online]. Available at: https://www.designingbuildings.co.uk/wiki/Information_management

Arifin, S. R. M. (2018). Ethical considerations in qualitative study. *International Journal of Care Scholars, 1*(2), 30–33.

Arslan, N., & Tathdil, H. (2012). Defining and measuring competitiveness: A comparative analysis of Turkey with 11 potential rivals. *International Journal of Basic and Applied Sciences, 12*(2), 31–43.

Arvantis, S. (2003). *Information technology, workplace organisation, human capital and firm productivity: Evidence for the Swiss economy.* Working Paper 74, Swiss Federal Institute of Technology Zurich.

Assaf, S. A., & Al-Hejji, S. (2006). Causes of delay in large construction projects. *International Journal of Project Management, 24*(4), 349–357.

Barlett, J. E., Kotrlik, J. W., & Higgins, C. C. (2001). Organizational research: Determining appropriate sample size in survey research. *Information Technology, Learning, and Performance Journal, 19*(1), 1–23.

Bartlett, K. R. (2001). A relationship between training and organizational commitment: A study in the health care field. *Human Resource Development Quarterly, 12*(4), 335–352.

Bates, M. J. (2006). Fundamental forms of information. *Journal of the American Society for Information Science and Technology, 57*(8), 1033–1045.

Begman, O., Tucker, S., Beyth-Marom, R., Cutrell, E., & Whittaker, S. (2009). *It is not that important: Demoting personal information of low subjective importance using Gray Area* (pp. 269–278).

Bergman, O., Gradovitch, N., Barllan, J., & Beyth-Maron, R. (2013). Tagging Personal information: A contrast between attitudes and behaviour. *Proceedings of the American Society for Information Science and Technology, 50*(1), 1–8.

Betheway, A. (2014). *Investing in information: The information management body of knowledge.* Springer International Publishing. https://doi.org/10.1007/978-3-319-11909-0.pp.25-34

Betts, M., Clark, A., & Ofori, G. (1999). The importance of new approaches to management. In *Strategic management of IT in construction* (pp. 3–13).

Beynon-Davies, P. (2006). Information: In-form to perform. In: *Significance* (pp. 174–201). Palgrave Macmillan. https://doi.org/10.1057/9780230295025

Bhargav-Spantzel, A., Squicciarini, A. C., & Bertino, E. (2006). Establishing and protecting digital identity in federation systems. *Journal of Computer Security, 14*(3), 269–300.

Bolívar-Ramos, M. T., García-Morales, V. J., & García-Sánchez, E. (2012). Technological distinctive competencies and organizational learning: Effects on organizational innovation to improve firm performance. *Journal of Engineering and Technology Management—JET-M, 29*(3), 331–337.

Bolumole, Y., Knemeyer, A., Michael, A., & Douglas, M. (2003). The customer service management process. *The International Journal of Logistics Management, 4*, 15–31.

Borner, K., Chen, C., & Boyack, W. K. (2003). Visualising knowledge domains. *Annual Review of Information Science and Technology, 37*(1), 175–255.

Brace, I. (2018). *Questionnaire design: How to plan, structure and write survey material for effective market research.* Kogan Page Publishers.

Brynolfsson, E., & Hitt, L. (2000). Information technology, organisational transformation and business performance. *Journal of Economic Perspectives, 14*(4), 23–48.

Cantarello, S., Martini, A., & Nosella, A. (2013). A Multi-level model for organisation: Ambidexterity in the search phase of the innovation process. *Creativity and Innovation Management, 21*(1), 28–48.

Charef, R., Alaka, H., & Emmitt, S. (2018). Beyond the third dimension of BIM: A systematic review of literature and assessment of professional views. *Journal of Building Engineering, 19*, 242–257. https://doi.org/10.1016/J.JOBE.2018.04.028

Cheah, J., Thurasamy, R., Memon, M. A., Chuah, F., & Ting, H. (2020). Multigroup analysis using smart PLS: Step by step guidelines for business research. *Asian Journal of Business Research, 10*(3), 1–15.

Colomina, I., Blázquez, M., Molina, P., Parés, M., & Wis, M. (2008). Towards a new paradigm for high-resolution low-cost photogrammetry and remote sensing. *Internstionsl Archiving Photogrammetry Remote Sensory Spatial Information Scienence, 37*(3), 1201–1206.

Connelly, L. M. (2014). Ethical considerations in research studies. *Medsurg Nursing, 23*(1), 54–56.

Currie, W. L. (2004). The organising vision of application service: A process-oriented analysis. *Information and Organisation, 14*, 237–267.

D4H. (2021). *Information management: Challenges in an an emergency response.* [Online]. Available at: www.d4htechnologies.com/blog/6-information-management-challenges-in-an-emergency-response

Dahiru, T. (2008). P-value, a true test of statistical significance? A cautionary note. *Annals of Ibadan Postgraduate Medicine, 6*(1), 21–26.

Dawood, N., Akinsola, A., & Hobbs, B. (2002). Development of automated communication of system for managing site information using internet technology. *Automation in Construction, 11*(5), 557–572.

De Pelsmacker, P., Muller, M. L., Viviers, W., Saayman, A., & Cuyvers, L. (2005). Competitive intelligence practices of South African and Belgian exporters. *Marketing Intelligence and Planning, 23*(6), 606–620.

Delgado, C. D., Clayton, P., O'Brien, W. J., Seepersad, C., Juenger, M., Ferron, R., & Salamone, S. (2018). Applications of additive manufacturing in the construction industry: A forward-looking review. *Automation in Construction Structure, 89*, 110–119. https://doi.org/10.1016/J.AUTCON.2017.12.031

Dou, H., Dou, M., & Manullang, S. D. (2005). The magic triangle—How to develop and apply competitive intelligence in developing countries. *Journal of Information Systems*, 1–7.

Dunamis, S., Cutrell, E., Cardiz, J., Jancke, G., Sarin, R., & Robbins D. (2003). *Suff I've seen: A system for personal information retreival and re-use* (pp. 72–79).

Dunne, C. (2011). The place of the literature review in grounded theory research. *International Journal of Social Research Methodology, 14*(2), 111–124.

Dwivedi, Y. K., Hughes, L., Ismagilova, E., Aarts, G., Coombs, C., Crick, T., Duan, Y., Dwivedi, R., Edwards, J., Eirug, A., Galanos, V., Janssen, M., et al. (2019). Artificial intelligence: Multidisciplinary perspectives on emerging challenges, oppourtunities and agenda for research, practice and policy. *International Journal of Information Management*, 1–47. https://doi.org/10.1016/j.ijinfomgt.2019.08.002

Eastern, C. M., Sacks, R., & Lee, G. (2002). *Strategies for realising the benefits of 3D integrated modeling of buildings for the AEC Industry* (pp. 9–14).

Eiselen, S. J. (2007). A basic mathematical skills test as predictor performance of tertiary level. *South African Journal of Higher Education, 21*(1), 38–49.

Fabrigar, L. R., & Wegener, D. T. (2011). Requirements for and decision in choosing exploratory common factor analysis. *Exploratory factor analysis: Understanding statistics* (pp. 19–38). Oxford University Press, Incorporated.

Fate, P. N. (2017). *Narrative knots/symbolic seduction: Toward a biopoetics of second-order symbolism in the storytelling art*. Doctoral Dissertation, University of California, Santa Barbara.

Fazli, A., Fathi, S., Enferadi, M. A., Fazli, M., & Fathi, B. (2014). Appraising effectiveness of Building Information Management (BIM) in project management. *Proceed in Technology, 16*, 1116–1125.

Frohne, P. T. (2008). Computer-aided logistics suppourt (CALS)/continous acquisition and life-cycle suppourt/(CALS)/Computer-aided acquisition and logistics support (CALS). In: *Quantitative measurements for logistics* (Vol. 360, pp. 252–266). GlobalSpec Engineering.

Gangatharan, S. (2015). Activities of project management in construction. In: *Project management in construction* (pp. 155–170). Blackwell Science.

Garelli, S. (2006). Competitiveness of nations: The fundamentals. In*IMD world competitiveness yearbook* (pp. 702–713).

Garner, W. R. (2014). *The processing of information and structure*. Psychology Press, Taylor & Francis Group.

Gasking, J., Godfrey, S., & Vance, A. (2018). Successful system use: It is not just what you are, but what you do. *AIS Transactions on Human-Computer Interaction, 10*(2), 57–81.

Nieezen, G., Eslambolchilar, P., & Thimbleby, H. (2016). Open-source hardware for medical devices. *BMJ Innovations*, 78–83.

Goertzen, M. J. (2017). Introduction to quantitative research and data. *Library Technology Reports, 53*(4), 12–18.

GoGrid. (2021). *GoGrid.* [Online]. Available at: https://www.idc.com/research/cloudcomputing/index.jsp

Gorla, N., Somers, T. M., & Wong, B. (2010). Organisational impact of system quality, information quality and service quality. *Journal of Strategic Information Systems, 19*(3), 207–228.

Gottwald, G. A., & Melbourne, I. (2008). Comment on reliability of the 0-test for chaos. *Physical Review, 77*(2), 028201–028211.

Hagell, P. (2004). Testing rating scale undimensionality using the principal component analysis (PCA)/t-test protocol with Rasch model: The primary of theory over statistics. *Open Journal of Statistics, 4*(6), 456–465.

Hair, J. F., Ringle, C. M., & Sarstedt, M. (2011). PLS-SEM: Indeed a silver bullet. *Journal of Marketing Theory and Practice, 19*(2), 139–152.

Hair, J. F., Sarstedt, M., Pieper, T. M., & Ringle, C. M. (2012). Applications of partial least squares path modeling in management journals: A review of past practices and recommendations for future applications. *Long Range Planning, 45*(5–6), 320–340.

Hair, J., Matthews, L. M., Matthewa, R. L., & Sarstedt, M. (2017). PLS-SEm or SEM: Updated guidelines on which method to use. *International Journal of Multivariate Data Analysis, 1*(2), 107–123.

Harris Milton, R. A. (2002). Organisation science. *Management Science, 48*(7), 852–865. https://doi.org/10.1287/mnsc.48.7.852.281

Hasselbring, W. (2000). Information system integration. *Communications of the Association for Computing Machinery., 43*(6), 32–38.

Hendrickson, R. M., Lane, J. E., Harris, J. T., & Dormon, R. H. (2017). *Academic leadership and governance of higher education: A guide for trustees* (pp. 1–418). Stylus Publishing.

Henseler, J., Ringle, C. M., & Sartedt, M. (2015). A new criterion for assessing discriminant validity in variance-based structural equation modelling. *Journal of the Marketing Science, 43*(1), 115–135.

Hideaki, M. (2012). *Institutional change and its economic consequenece in Japan: The bright and dark sides of hybridisation* (pp. 1–45). The Research Institute of Economy, Trade, and Industry.

Hillebrandt, P. M. (2000). *Economic theory and construction industry.* Palgrave Macmillan.

Holler, J., Tsiatsis, V., Catherine, M., Karnouskos, S., Avesand, S., & Boyles, D. (2014). *Internet of things: Introduction to a new age of intelligence.* Academic Press, Elsevier Limited.

Hu, G., Li-Xin, C., & Jian, L. (2008). Second order approximation of tool envelope surface for 5-axia machining with single point contact. *Computer Aided Design, 40*, 604–615.

Hu, J., Tung, W., Gao, J., & Cao, Y. (2005). Reliability of the 0–1 test for chaos. *Physical Review E, 72*(5), 056207-1–056207-5.

Huges, T. P. (2008). MIT architecture and values: Gehry's Stata and Holl's Simmons. *History and Technology, 24*(3), 207–220. https://doi.org/10.1080/034687503134749

Hugo, J. W. (2020). *Construction outlook for South Africa.* The Immediate Outlook.

Hyari, K. H. (2005). Introduction to construction industry. *Construction Project Management.*

IDC. (2021). *Cloud computing.* [Online]. Available at: https://www.idc.com/research/cloudcomputing/index.jsp

Ifan, H. K., Dou, J. M., Manullang, S., & Dou, H. (2004). Developing competitive technical intelligence in Indonesia. *Technovation, 24*, 995–999.

Jackson, D., Gillaspy, J., & Purc-Stephenson, R. (2009). Reporting practices in confirmatory factor analysis: An overview and some recommendation. *American Psychological Association, 14*(1), 6–23. https://doi.org/10.1037/a0014694

Jansen, D., & Warren K. (2021). *What (exactly) is research methodology? A plain-language explanation & definition (with examples).* [Online]. Available at: https://gradcoach.com/Whatisresearchmethodology

Jasch, C. (2006). How to perform an environmental management cost assesment in one day. *Journal of Cleaner Production*, 1194–1213.

Jay, R. (2012). The future of organisation design. *Journal of Organisation Design, 1*, 3–6.

Joseph, J., & Gaba, V. (2009). Organisational structure, information processing and decision making: A retrospect and roadmap for research. *Academy of Management Annals, Forthcoming*, 1–91.

Joseph, M., & Stone, G. (2003). An empirical evaluation of US bank customer perceptions of the impact of technology on service delivery in the banking sector. *International Journal of Retail & Distribution Management, 31*(4), 190–202. https://doi.org/10.1108/09590550310469185

Jung, Y. S., & Joo, M. H. (2011). Patterns and trends in cost control practice of Korean general contractors. *Korean Journal of Construction Engineering and Management, 12*(4), 79–87.

Jung, W., & Lee, G. (2015). The status of BIM adoption on six continents. *International Journal of Civil and Environmental Engineering, 9*(5), 512–516.

Kamel, M. N. (1995). *Applications of the continous acquisition and life-cycle suppourt (CALS)* (pp. 1–92). Naval Pstgraduate School.

Kasuya, E. (2019). On the use of R and R squared in correlation and regression. *Eclogical Research, 34*(1), 235–236. https://doi.org/10.1111/144-0-1703.1011

Katzenbach, J. R., & Smith, D. K. (2005). *The discipline of teams: Harvard Business Review* (pp. 111–120). Harvard Business Press.

Kikoski, C. K., & Kikoski, C. K. (2004). *The inquiring organisation: Tacit knowledge conversation and knowledge creation: Skills for 21st century organisations*. Praeger.

Kothari, C. R. (2004). *Research methodology: Method and technique* (2nd ed.). New age International (P) Limited, Publishers. ISBN (13): 978-81-224-2488-1

Kouider, T., Paterson, G., & Thomson, C. (2007). *Bim as a viable collaborative working tool: A case study* (pp. 57–68).

Kristensen, T. S. (2010). A questionnaire is more than a questionnaire. *Scandinavian Journal of Public Health, 38*(2), 149–155.

Laudon, K. C. (2006). *Management information systems: Managing the digital firm*. Pearson/ Prentice Hall. ISBN: 0132382350.

Lawal, G. (2007). Corruption and development in Africa: Challenges for political and economic change. *Humanity and Social Science Journal, 2*(1), 1–7.

Lederer, A. L., & Sethi, V. (2009). Key prescriptions for strategic information systems planning. *Journal of Management Information Systems, 13*(1), 35–62.

Li, J., & Yang, H. (2016). *A research on development of construction industrialization based on BIM technology under the background of Industry 4.0* (pp. 1–8).

Lietz, P. (2010). Research into questionnaire design: A summary of the literature. *International Journal of Market Research, 52*(2), 249–272.

Love, P. E. D., Irani, Z., Li, H., Chang, E. W. L., & Tse, R. Y. C. (2001). An Analysis of the barriers to implementing e-commerce in small-medium sized construction contractors in the state of Victoria, Australlia. *Construction Innovation, 1*, 31–41.

Lvenendonk, M. (2017). Management Information Systems (MIS): Definition and how it works. *Cleverism*, 1–12.

M., F. (2021). Stages of an information management implementation: Data management implementation. *Inetsoft Open Standard Implementation*, 1–3.

Matsunga, M. (2010). How to factor analyse your data right: Do's, don'ts and how-to's. *International Journal of Psychological Research, 3*(1), 97–110.

Mattews, C. E., Hadstromer, M., Pober, D. M., & Bowles, H. R. (2013). Best practices for using physical activity monitors in population-based research. *Medical Science Sports Exercise*, 1–15.

McConnel, V., & Wiley, K. (2010). Infill development: Perspectives and evidence from economics and planning. *Discussion Paper* (pp. 1–16).

McKim, B. R., & Saucier, P. R. (2011). An excel-based mean weighted discrepancy score calculator. *Journal of Extension, 49*(2), 39–42.

Melvin, N., Kraemer, K., & Gurbaxani, V. (2004). Information technology and organisational performance: An integrative model of IT business value. *MIS Quarterly, 28*(2), 283–322.

Moreau Karen, A., & Back, W. E. (2000). Improving design process with information management. *Automation in Construction, 10*(10), 127–140.

Muller, S., & Stegmaier, J. (2017). Why is there resistance to works councils in Germany? An economic perspective. *IWH Discussion Papers, Leibniz Institute for Economic Research (IWH), Halle* (pp. 1–39).

Nair, C. S., & Adams, P. (2009). Survey platforms: A factor influencing online survey delivery and response rate. *Quality in Higher Education, 15*(3), 291–296.

Nathan, K. (2016). *Construction information*. [Online]. Available at: construtiontuts.com/ krishanathan

Nathan, K. (2017). *Construction industry overview*. Report Linker.

Otasowie, K., Aigbavboa, C., Ikuabe, M., Adekunle, P., Oke A., & Adekunle, S. (2023). Barriers to the adoption of building information modeling in quantity surveying practice in South Africa. In *Proceedings of the 31st Annual Conference of the International Group for Lean Construction (IGLC31)* (pp. 41–49). https://doi.org/10.24928/2023/0137

Otasowie, K., Aigbavboa, C., Ikuabe, M., & Adekunle, P. (2024). Big data adoption in construction and demolition waste management: Prevailing challenges in developing nations. In K. Arai (Ed.), *Proceedings of the Future Technologies Conference (FTC) 2024, Volume 4. FTC 2024*. Lecture Notes in Networks and Systems (Vol. 1157). Springer. https://doi.org/10.1007/978-3-031-73128-0_42

Otasowie, K., Aigbavboa, C., Oke, A., Adekunle, P. (2024). Influencing factors for blockchain adoption in construction project logistics. In D. Li, P. X. W. Zou, J. Yuan, Q. Wang, Y. Peng (Eds.), *Proceedings of the 28th International Symposium on Advancement of Construction Management and Real Estate. CRIOCM 2023*. Lecture Notes in Operations Research. Springer. https://doi.org/10.1007/978-981-97-1949-5_130

Oyewobi, L. (2014). *Modelling performance differentials in large construction organisations in South Africa*. University of Capetown.

Page, M. (2012). Top 6 construction project challenges. *Page Group Management Advice*, 1–2.

Peansupap, V., & Walker, D. (2005). Factors enabling information and communication technology diffusion and actual implementation in construction organisations. *Electronic Journal of Information Technology in Construction, 10*, 193–218.

Pellissier, R. (2011). Understanding the use of strategic intelligence as a strategic management tool in the long-term insurance industry in South Africa. *South African Journal of Information Management, 13*(1), 1–13.

Buneman, P., Khanna, S., Tajima, K., & Tan, W.-C. (2004). Archiving scientific data. *ACM Transactions on Database Systems (TODS), 29*(1), 36–37.

Raghu, T. S., & Ajay, V. (2007). A business process context for knowledge management. *Decision Support System, 43*, 1062–1079.

Renz, A., & Solas, M. Z. (2016). *Shaping the future of construction: A breakthrough in mindset and technology* (pp. 1–22).

Rezgui, Y., Coorper, G., & Brandon, P. (1998). Information management in a collaborative multi-actor environment: The COMMIT approach. *Journal of Computing in Civil Engineering, 12*(3), 136–145. ISSN 0887-3801.

Rezgui, Y., & Zarli, A. (2006). Paving the way to the vision of digital construction: A strategic roadmap. *Journal of Construction Engineering Management, 132*(7), 767–776.

Riccio, E. L. (2004). Strategic information management: Challenges and strategies in managing information systems. *Journal of Information System and Technology Management, 1*.

Ryan, B. D., & Weber, R. (2007). Valuing new development in distressed urban neighborhoods. *Journal of the American Planning Association*, 100–111.

Ryu, S. (2013). Predictive analytics: The power to predict who will click, buy, lie or die. *Healthcare Informatics Research*, 63–65.

Sadeghi, A. R., Wachsmann, C., & Waidner, M. (2015). *Security and privacy challenges in industrial internet of things* (pp. 1–6). IEEE.

Saley, A., & Bista, K. (2017). Examining the factors impacting online survey response rates in educational research: Perceptions of graduate students. *Online Submission, 13*(2), 63–74.

Sauder, J.,Hilgemann, E., Bienstock, B., & Parness, A. (2017). *An automaton rover enabling long duration in-situ science in extreme environments* (pp. 1–10). IEEE. https://doi.org/10.1109/AERO.2017.7943829

Schnuch, A., Aberer, W., Agathos, M., Brasch, J., Frosch, P. J., Fuchs, T., & Richter, G. (2001). Leithinien der Deutschen Dermatologischen Gesellschaft (DDG) Zur Durchfuhrung des Epikutantests mit kontaklaergreen. *Der Hautarzt, 52*(10), 864–866.

Schuller, B., & Lidbom, M. (2009). Competitiveness of nations in the global economy. Is Europe internationally competitive? *Economics and Management, 14*, 934–939.

Sewdass, N., & Toit, A. D. (2014). Current state of competitive intelligence in South Africa. *International Journal of Information Management, 34*(2), 185–190.

Shirky, C. (2021). *Shirky*. [Online]. Available at: http://www.shirky.com/writings/ontology-overrated

Simon, B. (2017). Information management strategies: From punch cards to data warehouses, and looking to the future with big data and artificial intelligence. *SmartSheet*, 1–12.

Storck, J., & Hill, P. A. (2000). Knowledge diffusion through strategic communities. *Sloan Management Review, 41*(2), 63–74.

Strauss, A. C., & Du Toit, A. S. A. (2010). Competitive intelligence skills needed to enhance South Africa's competitiveness. *Aslib Proceedings: New Information Perspectives, 62*(3), 302–320.

Sukamolson, S. (2007). Fundamentals of quantitative research. *Language Institute Chulalongkorn University, 1*(3), 2007.

Techopedia. (2021). *Information Management (IM).* [Online]. Available at: http://techopedia.com/definition/200012/informationmanagement(im)

Tevan, J., Dunamis, T., & Horvitz, E. (2005). *Personalising search via automated analysis of interests and activities* (pp. 449–456).

Tombesi, P., & Whyte, J. (2011). Challenges of design management in construction. In: *The handbook of design management* (pp. 202–213).

Tumi, S. A. H., Omran, A., & Pakir, A. H. K. (2009). *Causes of delay in construction industry in Libya* (pp. 265–272).

Turk, Z., & Klinc, R. (2017). Potentials of blockchain technology for construction management. In *Proceeding Engineering, Creative Construction Conference.*

Tutupary, S. E., & Idianto, L. A. (2014). Benefits of management information system on the effectiveness and efficiency of the online business. *Journal of Business and Management, 3*(8), 835–849.

Umble, E. J., Haft, R. R., & Umble, M. M. (2003). Enterprise resource planning: Implementation procedures and critical success factors. *European Journal of Operational Research, 146*, 241–257.

Urbach, N., & Ahlemann, F. (2010). Structural equation modelling in information system research using partial least squares. *Journal of Information Technology Theory and Application, 11*, 5–40.

Vachara, P., & Derek, W. (2005). Factors enabling information and communication technology diffusion and actual implementation in construction organisations. *Itcon, 10*, 193–219.

Waheeduzzaman, A. N. (2002). Competitiveness, human development and inequality: A cross-national comparative inquiry. *Competitiveness Review: An International Business Journal, 12*(2), 13–17.

Wahlberg, J. E. (2001). Patch testing in Rycroft RJG. *Textbook of contact dermatitis* (pp. 435–468). Springer.

Walker, T. D. H. (2003). Implications of human capital issues. *Procurement strategies: A relationship based approach* (pp. 258–295). Blackwell Publishing.

Walrath, D. E., Turner, P., & Bruzek, J. (2004). Reliability test of the visual assessment of cranial traits for sex determination. *American Journal of Physical Anthropology, 125*(2), 132–137.

Walther, J. B., Van der Heicle, B., Kim, S., Westerman, D., & Tom-tang, S. (2008). The role of friends 'appearance and behaviour on evaluations of individuals on Facebook': Are we known by the company we keep? *Human Communication Research, 34*, 28–49.

Whyte, J., Bouchlagem, D., & Thorpe, T. (2002). IT implementation in the construction organisation. *Engineering Construction and Architectural Management, 9*(5–6), 371–377.

Wikforss, O., & Lofgren, A. (2007). Rethinking communication in construction. *Journal of Information Technology in Construction (ITcon), 12*(23), 337–346.

Winch, H. (2008). Pimp my respository. In*Drille Open Respositories Conference* (pp. 36–37).

Winter, L., Parks, S. M., & Diamoond, J. J. (2010). Ask a different question, get a different answer: Why living wills are poor guides to care preference to end of life. *Journal of Palliative Medicine, 13*(5), 567–572.

Wollam, M. (2009). *Construction management: The power of teamwork in construction* (pp. 1–9). Wollam Construction.

Wong, K. K. (2013). Partial least square structural equation modelling (PLS-SEM) technique using SmartPLS. *Marketing Bulletin, 24*(1), 1–32.

Zahirovich-Herbert, V., & Gilber, K. M. (2014). The effect of new residential construction on housing prices. *Journal Housing Economics, 26*, 1–18.

Zaiţ, A., & Bertea, P. E. (2011). Methods for testing discriminant validity. *Management & Marketing Journal, 9*(2), 217–224.

Zhao, L., Tiann, L., Cai, T., Clagget, B., & Wei, L. (2013). Effectively selecting a target population for a future comparative study. *Journal of the American Statistical Association, 108*(502), 527–539.

MIX
Papier aus verantwortungsvollen Quellen
Paper from responsible sources
FSC® C105338

If you have any concerns about our products,
you can contact us on
ProductSafety@springernature.com

In case Publisher is established outside the EU,
the EU authorized representative is:
Springer Nature Customer Service Center GmbH
Europaplatz 3, 69115 Heidelberg, Germany

Printed by Libri Plureos GmbH
in Hamburg, Germany